MOLECULAR APPROACHES TO IMPROVING FOOD QUALITY AND SAFETY

Edited by

Deepak Bhatnagar

and

Thomas E. Cleveland

An **avi** Book
Published by Van Nostrand Reinhold
New York

An AVI Book
(AVI is an imprint of Van Nostrand Reinhold)

Copyright © 1992 by Van Nostrand Reinhold
Softcover reprint of the hardcover 1st edition 1992

Library of Congress Catalog Card Number 91-43120

ISBN 978-1-4684-8072-6 ISBN 978-1-4684-8070-2 (eBook)
DOI 10.1007/978-1-4684-8070-2

Van Nostrand Reinhold
115 Fifth Avenue
New York, New York 10003

Chapman and Hall
2-6 Boundary Row
London, SE1 8HN, England

Thomas Nelson Australia
102 Dodds Street
South Melbourne 3205
Victoria, Australia

Nelson Canada
120 Birchmount Road
Scarborough, Ontario MIK 5G4, Canada

16 15 14 13 12 11 10 9 8 7 6 5 4 3 2 1

Library of Congress Cataloging-in-Publication Data
Molecular approaches to improving food quality and safety/edited by
    Deepak Bhatnagar and Thomas E. Cleveland.
        p. cm.
    Includes bibliographical references and index.

    1. Food—Biotechnology. I. Bhatnagar, Deepak, 1949–
II. Cleveland, Thomas E.
TP248.65.F66M65 1992
664—dc20   91–43120

Contents

Foreword, v
Preface, vii
Contributors, ix

1. Enhancing the Nutritional Quality of Crop Plants: Design,
   Construction, and Expression of an Artificial Plant
   Storage Protein Gene   1
   JH. Kim, S. Cetiner, and J. M. Jaynes

2. Deamidation and Phosphorylation to Improve Protein
   Functionality in Foods   37
   F. F. Shih, J. S. Hamada, and W. E. Marshall

3. Natural Enzyme and Biocontrol Methods for Improving
   Fruits and Fruit Quality   61
   E. A. Baldwin and R. A. Baker

4. Safety Evaluation of Food Enzymes from Genetically
   Engineered Organisms   83
   N. W. Zeman and W. Martin Teague

5. Toward the Genetic Engineering of Disease Resistance
   in Plants: The Transfer of Pea Genes to Potatoes   99
   L. A. Hadwiger

6. Automated System for Microbial Screening/Breeding   113
   T. Okuda and H. Tabuchi

7. DNA Probes for the Identification of Pathogenic
   Foodborne Bacteria and Viruses   151
   K. A. Lampel, P. Feng, and W. E. Hill

8. Rapid Methods for the Detection of *Listeria*   189
   J. Gavalchin, K. Landy, and C. A. Batt

9. Molecular Strategies for Reducing Aflatoxin Levels
   in Crops before Harvest   205
   T. E. Cleveland and D. Bhatnagar

10. Molecular Strategies for Improving the Quality of Muscle
    Food Products   229
    A. M. Spanier and P. B. Johnsen

Index, 242

Foreword

Conventional food safety and quality research have traditionally dealt with the whole organism or food product, either plant or animal, or with the microorganisms that associate with these food-producing plants and animals at various stages of growth, development, and maturation. However, conventional research methods no longer are sufficient to bring about the improvements in quality and safety of foods that are demanded in today's marketplace by increasingly educated and sophisticated consumers.

Improved quality has generally been thought to mean (1) unblemished products, particularly fruits and vegetables, and (2) the desired functionality of protein, carbohydrates, and fats in grain oilseeds and the like to achieve the intended purpose in processing foods; for example, dough characteristics to make bread or pasta. However, it also means the year-round ready availability of nutritious and appealing choices in all food groups that can be quickly and easily prepared. This promotes optimum nutrition, particularly necessary in children, the elderly, and special-needs groups, and it allows health conscious consumers to match their calorie intake with individual needs to achieve and maintain desirable body weight.

Improved food safety is also increasing in complexity as we measure smaller amounts of both chemicals and microbes and understand their significance. Thus food safety now means (1) freedom from pathogenic microorganisms; (2) disease and pest resistance in plants to help assure absence of illegal amounts of residues; (3) freedom from toxic heavy metals and mycotoxins; (4) varieties of plant commodities, fruits, vegetables, and grains containing the lowest possible amounts of known deleterious natural constituents; and (5) rapid on-line assay methods requiring a minimum of equipment and skill for accurate use by producers, processors, and the various regulatory agencies. Meeting all these criteria is a large order, one where the traditional research methods need the assistance of the newer molecular techniques to bring about the desired precise changes in the commodity or food product.

The current volume is designed to provide a catalyst for a potential rapid expansion in development of new, higher quality and safe food products through use of biotechnology. The research described in each of the chapters provides a "cross section" of examples of how particular changes or improvements in food quality or safety can be brought about using recently acquired knowledge of the biochemistry and molecular biology of microbes and higher organisms. Examples include alteration of quality and quantity of proteins in plants by plant genetic engineering. This can change the basic genetic capability of the plant and lead to new crop lines that are resistant to pests (some of which produce potent toxins) or are improved with regard to their nutritional value or other characteristics such as fruit ripening. Examples are also provided of biotechnological methods for improvement of the quality of muscle foods by genetic manipulation of animals. Looking to the future, the vast resource of microbial genetic material is examined for sources of enzymes and potential biocontrol agents used in improving food quality and safety. Also, the recently available and powerful technology of using gene probes for identification of potentially dangerous microbes is described.

Finally, the editors of this book at Southern Regional Research Center of the Agricultural Research Service are using molecular approaches in their own research to eliminate a major food safety problem, that is, the threat of aflatoxin contamination of food crops. Some hopeful possibilities for controlling aflatoxin in preharvest crops are examined based on recent breakthroughs in understanding the aflatoxin biosynthetic process at the molecular level and in elucidating the mechanism(s) of how toxin-producing fungi successfully invade seed tissues. The possibility is explored of using new precision techniques to eliminate aflatoxin pathway genes to produce stable non-toxin-producing strains that could be introduced into ecosystems as biocompetitive agents without the hazard of subsequent transfer of genetic information for toxin production into the engineered strains. With the recent breakthroughs in molecular engineering of plants, the possibility of incorporating alien genes into new elite crop varieties, which could endow the seed with resistance to growth of aflatoxin producing fungi, is also examined.

The editors have used this vast experience in the selection of authors and topics for each of the chapters. The result is a book that demonstrates how dreams can be translated into realities in producing and assuring a high-quality, safe food supply. This volume will not only be an excellent reference for researchers but also serve as a landmark for the food industry in its desire to set future goals for improving food safety and quality through advances in biotechnology.

DR. JANE F. ROBENS
National Program Leader
Food Safety and Health
USDA, Agricultural Research Service

Preface

A high priority of food microbiologists is to provide, through fundamental and applied research, the means for solving technical food and agricultural problems by improving the quality and safety of food available to the consumer. The next decade appears to be one of enormous change in the food industry. One agent of that change will be the continued application of biotechnology to postharvest food processing, product enhancement, and the safety in the use of agricultural commodities. Biotechnology refers to the procedures of using microorganisms and biological systems or their products to produce useful substances or to improve existing species. These procedures are ultimately aimed at developing commercial processes and products. Biotechnology therefore provides economical and efficient solutions to the rising problems of increasing energy costs, pollution, and depletion of world renewable resources.

Biotechnology is not a "new" development. For many centuries human beings have used selected organisms for their products: yeast in bread, wine, and cheese making; other microbes for antibiotics and pharmaceuticals. However, in this "traditional" form of biotechnology, selection was based mostly on breeding of plants or animals for the desired trait. The "modern" approach to biotechnology (since 1960) involves recombinant DNA techniques, "gene jockeying," tissue culture and plant regeneration, and immunological studies, to name a few.

Modern biotechnology has been applied primarily to human health care where molecular manipulations of microorganisms have resulted in economically producing significant quantities of therapeutic substances, for example, human growth hormone, vaccines, and insulin, as well as diagnostic tools for disease-causing organisms. Molecular and cellular biotechnology has only recently found its use in agriculture for improvement of the quality of crop and livestock production, as well as in food and feed safety. Examples are development of genetically altered crops that protect themselves against microbial or insect pests or possess enhanced nutritional value; use of genetically modified microorganisms as biocontrol agents; utilization of microbes for efficient production of enzymes required for improved food quality; microbially derived animal growth hormones for increased meat and milk production or for improved quality of meat. Research has also yielded chemical and biochemical processes for enhancing food and feed usage. Biotechnology has also been applied to the development of gene probes for rapid identification of pathogenic foodborne organisms to improve food and feed safety. However, in an atmosphere of being overwhelmed by the potential benefits of these developments, the risks of agricultural biotechnology must be assessed and a careful evaluation of the safety of these products of genetic recombination must be carried out.

The current volume has been commissioned not only to provide a range of perspectives on the application of biotechnology in food and feed safety and quality to those outside the immediate biotechnology area, but also designed to provide future approaches in research and development for exploring the potential of commercialization of the products of sustained agriculture. Scientists who have contributed to this volume have presented a cross section of exciting information; scientific information highlighting key issues in agricultural biotechnology have been included in the expositions. The editors would like to thank the authors for their contributions, other scientists for their critical suggestions, and the editorial staff of Van Nostrand Reinhold for careful review of the material. Their efforts have resulted in a significant presentation.

Contributors

Robert A. Baker, USDA/Agricultural Research Service/Citrus and Subtropical Products Lab, Winter Haven, Florida

Elizabeth A. Baldwin, USDA/Agricultural Research Service/Citrus and Subtropical Products Lab, Winter Haven, Florida

Carl A. Batt, Department of Food Science, Cornell University, Ithaca, New York

Deepak Bhatnagar, USDA/Agricultural Research Service/Southern Regional Research Center, New Orleans, Louisiana

Selim Cetiner, Department of Horticulture, Cukurova University, Adana, Turkey

Thomas E. Cleveland, USDA/Agricultural Research Service/Southern Regional Research Center, New Orleans, Louisiana

Peter Feng, Division of Microbiology, Food and Drug Administration, Washington, D.C.

Jerrie Gavalchin, Department of Medicine, SUNY Health Science Center, Syracuse, New York

Lee A. Hadwiger, Department of Plant Pathology, Washington State University, Pullman

Jamel S. Hamada, USDA/Agricultural Research Service/Southern Regional Research Center, New Orleans, Louisiana

Walter E. Hill, Seafood Products Research Center, Food and Drug Administration, Bothell, Washington

Jesse M. Jaynes, Department of Biochemistry, Louisiana State University, Baton Rouge

Peter B. Johnsen, USDA/Agricultural Research Service/Southern Regional Research Center, New Orleans, Louisiana

JaeHo Kim, Department of Biochemistry, Louisiana State University, Baton Rouge

Keith A. Lampel, Division of Microbiology, Food and Drug Administration, Washington, D.C.

Katrina Landy, Department of Biochemistry, University of Ottawa, Canada

Wayne E. Marshall, USDA/Agricultural Research Service/Southern Regional Research Center, New Orleans, Louisiana

Toru Okuda, Department of Microbiology, Nippon Roche Research Center, Kamakura, Japan

Frederick F. Shih, USDA/Agricultural Research Service/Southern Regional Research Center, New Orleans, Louisiana

Arthur M. Spanier, USDA/Agricultural Research Service/Southern Regional Research Center, New Orleans, Louisiana

Hiroyoshi Tabuchi, Umetani Precision Company, Tokyo, Japan

W. Martin Teague, Enzyme Bio-Systems Ltd., Arlington Heights, Illinois

Nancy W. Zeman, Consultant, Chapel Hill, North Carolina

1

Enhancing the Nutritional Quality of Crop Plants: Design, Construction, and Expression of an Artificial Plant Storage Protein Gene

JaeHo Kim, Selim Cetiner,
and Jesse M. Jaynes

The composition of storage proteins, a major food reservoir for the developing seeds, determines the nutritional value of plants and grains that are used as foods for man and domestic animals. The amount of protein varies with genotype or cultivar, but in general, cereals contain 10% of the dry weight of the seed as protein, while in legumes, the protein content varies between 20% and 30% of the dry weight. In many seeds, storage proteins account for 50% or more of the total protein, and thus determine the protein quality of seeds. Each year the total world cereal harvest amounts to some 1700 million tons of grain (Keris et al. 1985). This harvest yields about 85 million tons of cereal storage proteins harvested each year and contributes about 55% of the total protein intake of humans.

With respect to human and animal nutrition, most seeds do not provide a balanced source of protein because of deficiencies in one or more of the essential amino acids in the storage proteins. Humans require eight amino acids from foods—isoleucine, leucine, lysine, methionine, phenylalanine, threonine, tryptophan, and valine—to maintain a balanced diet. Consumption of proteins of unbalanced composition of amino acids can lead to a malnourished state that is most often found in children in developing countries where plants are the major source of protein intake. Therefore, the development of a more nutritionally balanced protein for introduction into plants is of extreme importance.

Recently, many laboratories have attempted to improve the nutritional

1

quality of plant storage proteins by transferring heterologous storage protein genes from other plants (Pederson et al. 1986). The development of recombinant DNA technology and the agrobacterium-based vector system has made this approach possible. However, genes encoding storage proteins containing a more favorable amino acid balance do not exist in the genomes of major crop plants. Furthermore, modification of native storage proteins has met with difficulty because of their instability, low level of expression, and limited host range. One possible alternative would be the *de novo* design of a more nutritionally balanced protein that retains certain characteristics of the natural storage proteins of plants.

Our initial work described the use of small fragments of DNA that encoded spans of protein high in essential amino acids (Jaynes et al. 1986; Yang et al. 1989). Subsequently, the genes encoding these protein domains were cloned into an existing protein and the expression level of this modified protein determined in transgenic potato plants. However, because of some of the problems previously mentioned, the results were somewhat less than desirable (Yang et al. 1989).

There are at least two fundamental difficulties in achieving efficient expression of designed proteins. First, it is not yet known what stabilizes a protein against proteolytic breakdown and second, the mechanisms for folding of an amino acid sequence into a biologically stable tertiary structure have not yet been fully delineated. For the design of ASP 1 (artificial storage protein) we focused on construction of a physiologically stable, as well as a highly nutritious, storage-protein-like artificial protein. Based on what we have learned in the design and expression of lytic peptides for enhanced disease resistance in plants (Jaynes et al. 1992), it now seemed possible for us to design entirely artificial, stable, nutritionally significant proteins to improve the nutritive quality of plants.

AMINO ACID REQUIREMENTS OF HUMANS

The biosynthesis of amino acids from simpler precursors is a process vital to all forms of life because these amino acids are the building blocks of proteins. Organisms differ markedly with respect to their ability to synthesize amino acids. In fact, virtually all members of the animal kingdom are incapable of manufacturing some amino acids. There are 20 common amino acids that are utilized in the fabrication of proteins, and essential amino acids are those protein building blocks that cannot be synthesized by the animal. It is generally agreed that humans require 8 of the 20 common amino acids in their diet. Protein malnutrition can usually be ascribed to a diet that is deficient in one or more of the essential amino acids. A nutritionally adequate diet must include a minimum daily consumption of these amino acids (Fig. 1–1).

When diets are high in carbohydrates and low in protein over a protracted period, essential amino acid deficiencies result. The name given to this undernourished condition is *kwashiorkor,* which is an African word meaning "deposed child" (deposed from the mother's breast by a newborn sibling). This debilitating and malnourished state, characterized by a bloated stomach and reddish-orange discolored hair, is more often found in children than adults because of a greater need for essential amino acids during growth and development. In order for normal physical and mental maturation to occur, the previously mentioned daily source of essential amino acids is a requisite. Essential amino acid content, or protein quality, is as important a feature of the diet as total protein quantity or total calorie intake.

Some foods, such as milk, eggs, and meat, have very high nutritional values because they contain a disproportionately high level of essential amino acids. On the other hand, most foodstuffs obtained from plants possess a poor nutritional value because of their relatively low content of some

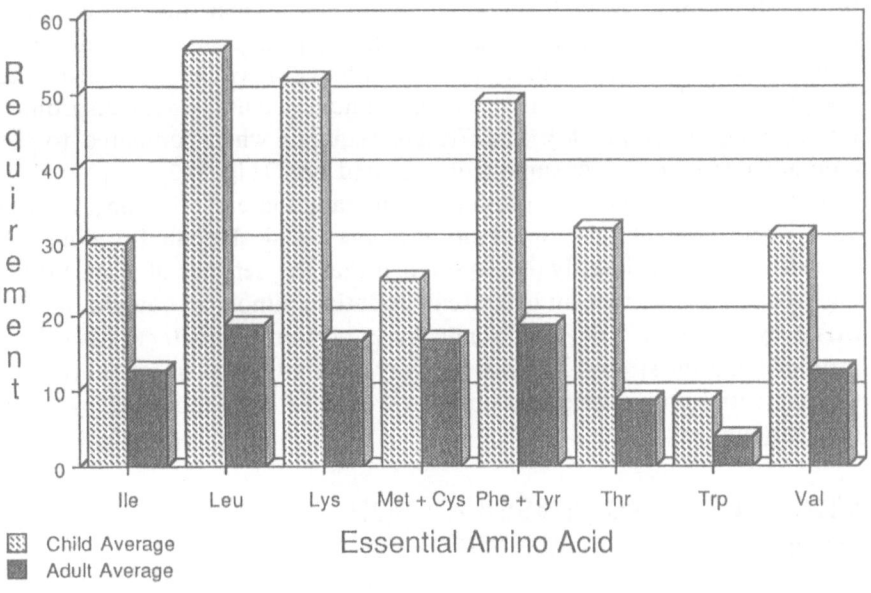

FIGURE 1-1.   The average essential amino acid requirement for both children and adults in milligrams per kilogram body weight. Note that children, on a per weight basis, require more of the essential amino acids than do adults. This difference indicates the importance of diet to normal physical and mental maturation.

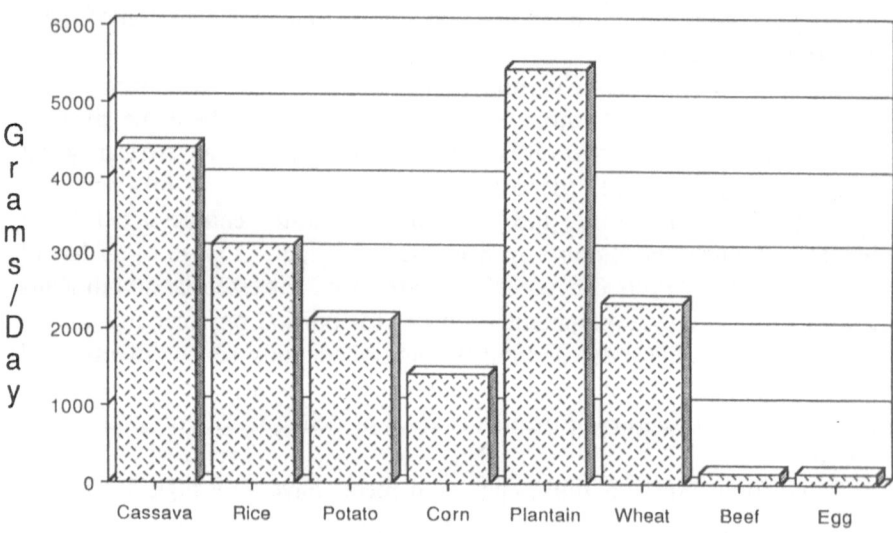

FIGURE 1-2.   The amount of foodstuffs that must be consumed in a gram per day in order to meet the minimum daily requirement of all essential amino acids.

or, in a few cases, all of the essential amino acids. Generally, the essential amino acids that are found to be most limiting in plants are isoleucine, lysine, methionine, threonine, and tryptophan (MLEAA). To satisfy the complete essential amino acid needs of the human child, a very disproportionate amount of plant foodstuffs are required when compared to the amounts necessary to consume from egg and beef (Fig. 1-2).

It has been difficult to produce significant increases in the essential amino acid content of crop plants utilizing classical plant breeding approaches. This is primarily due to the fact that the genetics of plant breeding is complex, and that an increase in essential amino acid content may be offset by a loss in other agronomically important characters. Also, it is probable that the storage proteins are very conserved in their structure, and their essential amino acid composition would be little modified by these conventional techniques.

STRUCTURE AND CLASSIFICATION OF NATURAL STORAGE PROTEINS

Seed storage proteins can be characterized by several main features (Pernollet and Mosse 1983):

• Their main function is to provide amino acids or nitrogen to the young seedling

- The general absence of any other known function
- Their peculiar amino acid composition in cereal and legume seeds
- Their localization within storage organelles called *protein bodies,* at least during seed development

Several classes of storage proteins are generally recognized based on their solubilities in different solvents. Proteins soluble in water are called *albumins;* proteins soluble in 5% saline, *globulins;* and proteins soluble in 70% ethanol, *prolamins.* The proteins that remain following these extractions are treated further with dilute acid or alkali, and are named *glutelins.* Most cereals contain primarily prolamin-type proteins and can be classified into different groups on the basis of the relative proportions of prolamins, glutelins, and globulins, and the subcellular location of these proteins in the mature seed. The first group corresponds to the Panicoideae subfamily, the second group the Triticeae tribe, and the last one to oat and rice storage proteins.

The principal members of the Panicoideae sub-family are maize, sorghum, and millet. Their major storage proteins are prolamins (50–60% of seed protein) and glutelins (35–40% of seed protein) (Pernollet and Mosse 1983). Prolamins are stored within protein bodies, but glutelins are located both inside and outside these organelles. The Triticeae tribe, which includes wheat, barley, and rye, differ from the Panicoideae mainly in storage protein localization and structure. In the starchy endosperm of the seeds belonging to this tribe, no protein bodies are left at maturity. Clusters of proteins are then deposited between starch granules, but are no longer surrounded by a membrane.

In legumes and most other dicots, the major storage proteins are salt-soluble globulins (80%) and prolamins (10–15%). Globulins can be divided into vicillins and legumins (Agros, Naravana, and Nielsen 1985), based on their sedimentation coefficient (7S/11S), oligomeric organization (trimeric/ hexameric), and polypeptide chain structure (single chain/disulphide-linked pair of chains). In the legume seed cotyledon, protein bodies are embedded between starch granules (Pernollet and Mosse 1983). They are membrane-bound organelles, a few microns in diameter, mainly filled with storage proteins and phytates. Besides storage proteins, protein bodies also contain other proteins, such as enzymes or lectins, although in lesser amounts.

The structure of soluble globulins was studied more than the insoluble prolamins and glutelins. Vicillin appears as a homo- or heterotrimer, sometimes able to associate into hexameric form. Soybean β-conglycin and French bean phaseolin (Bollini and Chrispeels 1978) are the structurally best known vicillins. Recently, the three-dimensional structure of phaseolin was determined by X-ray crystallographic analysis (Lawrence et al. 1990). However, unlike other vicillins, the phaseolin trimer can associate into a dode-camer (tetramer of trimer) below pH 4.5. Each polypeptide of the trimeric form comprises two structurally similiar units, each made up of a β-barrel and an α-helical domain.

Glycinin, the soybean legumin, has a quaternary structure that was sug-gested by Badley et al. (1975) to be 12 subunits packed in two identical hexagons. In general, the legumin molecule is a polymer formed by the association of six monomers. Each monomer consists of two subunits, acidic and basic. Sometimes, these subunits are associated by disulfide link-ages. On the other hand, arachin, the peanut legumin, was found to consist of different kinds of subunits. The arachin hexamer association does not need different kinds of subunits, which suggests that the subunits have a very similar structure.

The most studied storage proteins, in terms of structure, are the corn prolamines called *zeins*. These proteins perform no known enzymatic func-tion. Three types of zeins (α, β, and γ) (Esen 1986) are synthesized on rough endoplasmic reticulum and aggregate within this membrane as protein bod-ies. The zein protein readily self-associates to form protein bodies and is insoluble in water even in low concentrations of salt. The presence of all types of zeins is not necessary for the formation of a protein body since a single type of zein can aggregate into a dense structure and is generally found at the surface of protein bodies (Lending et al. 1988; Wallace et al. 1988). The mechanism responsible for protein body formation is thought to involve hydrophobic and weak polar interactions between individual zein molecules (Wallace et al. 1988; Agros et al. 1982), while they require a high amoung of ethanol in aqueous systems to maintain their strict molecular conformation (Agros et al. 1982).

Circular dichroic measurements, amino acid sequence analysis, and elec-tron microscopy of a zein protein suggests that zein secondary structure is primarily helical with nine adjacent, topologically antiparallel helices clus-tered within a distorted cylinder (Agros et al. 1982; Larkins 1983; Larkins et al. 1984). Polar and hydrophobic residues appropriately distributed along the helical surfaces allow intra- and intermolecular hydrogen bonds and van der Waals interactions among neighboring helices, such that rod-shaped zein molecules could aggregate and then stack through glutamate interactions at the cylindrical caps. Because of this structure, zein is much less soluble under physiological conditions than the globulin phaseolin, and precipitation of insoluble zein in the tightly packed protein body may make them less available for proteolytic degradation (Greenwood and Chrispeels 1985).

The storage protein structures are adapted to a maximal packing within protein bodies (Pernollet and Mosse 1983). Maximal packing is achieved in at least one of two ways. The folding of the polypeptide chain may favor the maximal packing of amino acids within the protein molecule, or the compacting of proteins is increased by the formation of closely packed quaternary structure. High degrees of polymerization can be observed in pearl millet pennisetin (Pernollet and Mosse 1983) or zein (Lending et al. 1988; Wallace et al. 1988). Also, wheat prolamins and glutelin associate into aggregates arising in the formation of insoluble gluten. These insoluble forms of protein deposits are osmotically inactive and stable during the long period of storage between the time of seed maturation and germination.

REGULATION OF STORAGE PROTEIN GENES

All storage proteins that have been investigated are encoded by multigene families (Bartels and Tompson 1983; Crouch et al. 1983; Forde et al. 1985; Kasarda et al. 1984; Lycett et al. 1985; Rafalski et al. 1984; Slightom, Sun, and Hall 1983). The structure of these families varies, in some cases, because in wheat or barley, two major subgroups can be noted: the α- and γ-gliadins and the B- and C-hordeins, respectively (Forde et al. 1985; Kasarda et al. 1984; Rafalski et al. 1984). Within each subgroup, several subfamilies can be distinguished. Often short repeats account for at least part of the structure of the polypeptides. These repeats constitute links through which different subfamilies within the same species are related.

Storage protein genes, like most other plant genes characterized to date, are transcribed in a regulated rather than a constitutive fashion. Expression is frequently tissue-specific and/or temporally regulated. Cis-acting DNA sequences involved in developmental and/or tissue-specific regulation gene expression can be defined by introducing plant storage protein gene regulatory regions coupled to bacterial reporter genes (Twell and Ooms 1987; Wenzler et al. 1989, Marries et al. 1988; Chen, Pan, and Beachy 1988), or by introducing entire or dissected genes (Colot et al. 1987; Chen, Schuler, and Beachy 1986) into a transgenic environment. Unfortunately, a transformation system for the nutritionally important cereal species has not yet been well established. Therefore, most regulation mechanisms have been studied with transgenic dicot plants. However, there is increasing evidence that gene expression is controlled, at least partly, by the interaction between regulatory molecules and short sequences that are present in the 5' flanking region of the gene.

The regulatory sequences of potato storage protein were investigated using transgenic potato plants. A 2.5-kb 5' flanking DNA fragment containing the promoter and the patatin gene was used to construct a transcriptional fusion gene with chloramphenicol acetyl transferase (CAT) or the β-glucuronidase (GUS) gene (Twell and Ooms 1987; Wenzler et al. 1989).

When reintroduced into potato, these chimeric genes were expressed in tubers, but not in leaves, stems, or roots.

The expression pattern of storage protein genes of cereals is retained in tobacco, not only with respect to tissue but also to temporal expression. The 5' upstream regions of wheat glutenin genes possess regulatory sequences that determine endosperm-specific expression in transgenic tobacco (Colot et al. 1987). Deletion analysis of the low molecular weight (LMW) glutenin sequence indicated that sequences present between 326 base pairs (bp) and 160 bp upstream of the transcriptional start point are necessary to confer endosperm-specific expression. Furthermore, cis-acting elements determining the regulation of each gene in the cluster are recognized by the tobacco trans-acting factor; also, cis-acting elements directing expression of one gene do not affect the expression of neighboring genes. This was demonstrated by the transfer of a 17.1-kb soybean DNA containing a seed lectin gene with at least four nonseed protein genes to transgenic tobacco plants (Okamuro 1986). The genes in this cluster were expressed in a manner similar to that in soybean; that is, the lectin gene products accumulated in seeds, and the other genes were expressed in tobacco leaves, stems, and roots.

The expression of several DNA deletion mutants with a 257 bp 5' flanking sequence of the a'-conglycin gene indicates that this region contained enhancer-like elements (Chen et al. 1986). Only a low level of expression of the a' gene occurred in developing seeds of transgenic plants that contain the a' gene flanked by 159 nucleotides 5' of the transcriptional start site. However, a twentyfold increase in expression occurred when an additional 98 nucleotides of upstream sequence were included. The DNA sequence between 143 and 257 contained five repeats of the sequence AA(G)CCCA, and played a role in conferring tissue-specific and developmental regulation. The 35S promoter containing this sequence in different positions and different orientations could enhance the expression of the CAT gene by twenty-five- to fortyfold (Chen et al. 1988).

Trans-acting factors directly involved in storage protein gene regulation have not yet been reported. However, in some cases, the level of amino acids can control the expression of storage protein. Vegetative storage protein (VSP) gene expression in leaves, stems, and seed pods is closely related to whether these organs are currently a sink for nitrogen or a source for mobilized nitrogen for other organs (Staswick 1989). The leaves have a sensitive mechanism for detecting changes in sink demand of mobilizing reserves, and VSP gene expression can be rapidly adjusted accordingly. Sequestering excess amino acids in this way may prevent their accumulation to toxic levels.

GENETIC ENGINEERING USING AGROBACTERIUM TUMEFACIENS

One of the most significant recent advances in the area of plant molecular biology has been the development of the *Agrobacterium tumefaciens* Ti plasmid as a vector system for the transformation of plants. In nature, A. tumefaciens infects most dicotyledonous and some monocotyledonous plants by entry through wound sites. The bacteria bind to cells in the wound and are stimulated by phenolic compounds released from these cells to transfer a portion of their endogenous, 200-kb Ti plasmid into the plant cell (Weiler and Schroder 1987). The transferred portion of the Ti plasmid (T-DNA) becomes covalently integrated into the plant genome, where it directs the biosynthesis of phytohormones using enzymes that it encodes. The vir gene in the bacterial genome is known to be responsible for this process. In addition to vir gene products, directly repeating sequences of 25 bases called ''border'' sequences are essential, but only the right terminus has been shown to be used for T-DNA transfer and integration.

Expression of the T-DNA gene inside the plants results in the uncontrolled growth of these and surrounding cells, leading to formation of a gall (Weiler and Schroder 1987). Ti plasmids, from which these disease-producing genes have been removed or replaced, are referred to as *disarmed,* and can be used for the introduction of foreign genes into plants. The great size of the disarmed Ti plasmid and lack of unique restriction endonuclease sites prohibit direct cloning into the T-DNA. Instead, intermediate vectors such as pMON237 or pBI121 can be used to introduce genes into the Ti plasmid. Currently, two kinds of vector systems are available as intermediate vectors: cointegrating vectors and binary vectors. A cointegrating transformation vector must include a region of homology between the vector plasmid and the Ti plasmid. Once recombination occurs, the cointegrated plasmid is replicated by the Ti plasmid origin of replication. The cointegrate system, while more difficult to use, does offer advantages. Once the cointegrate has been formed, the plasmid is stable in agrobacterium.

A binary vector contains an origin of replication from a broad host-range plasmid instead of a region of homology with the Ti plasmid. Since the plasmid does not need to form a cointegrate, these plasmids are considerably easier to introduce into agrobacterium. The other advantage to binary vectors is that this vector can be introduced into any agrobacterium host containing any Ti or Ri plasmid, as long as the vir helper function is provided. Using these systems, the gene regulation mechanism of storage proteins has been elucidated.

IMPROVEMENT OF NUTRITIONAL QUALITIES OF PLANTS

The amino acid composition of the cereal endosperm protein is characterized by a high content of proline and glutamine, while the amount of essential amino acids, lysine and tryptophan in particular, is a limiting factor (Pernollet and Mosse 1983). In legumes, sulfur-containing amino acids such as methionine and cysteine are the major limiting essential amino acids for the efficient utilization of plant protein as animal or human food, while roots and tubers are deficient in almost all of the essential amino acids.

There has been a great deal of effort to overcome these amino acid limitations by breeding and selecting for more nutritionally balanced varieties. Plants have been mutated in hopes of recovering individuals with more nutritious storage proteins. Neither of these approaches has been very successful, although some naturally occurring and artificially produced mutants of cereals were shown to contain a more nutritionally balanced amino acid composition. These mutations cause a significant reduction in the amount of storage protein synthesized, and thereby result in a higher percentage of lysine in the seed; however, the softer kernels and low yield of such strains have limited their usefulness (Pernollet and Mosse 1983). The reduction in storage protein also causes the seeds to become more brittle; as a result, these seeds shatter more easily during storage. The lower levels of prolamin also result in flours with unfavorable functional properties that cause brittleness in the baked products (Pernollet and Mosse 1983). Thus, no satisfactory solution has yet been found for improving the amino acid composition of storage proteins.

One direct approach to this problem would be to modify the nucleotide sequence of genes encoding storage proteins so that they contain high levels of essential amino acids. To achieve this aim, several laboratories have tried to modify and express storage proteins in the host plants. Modified storage proteins have been created and expressed by changing their codon sequences. In vitro mutagenesis was used to supplement the sulfur amino acid codon content of a gene-encoding β-phaseolin, a *Phaseolus vulgaris* storage protein (Hoffman, Donaldson, and Herman 1988). The nutritional quality of β-phaseolin was increased by the insertion of 15 amino acids, six of which were methionine. The inserted peptide was essentially a duplication of a naturally occurring sequence found in the maize 15-kilodalton (kd) zein storage protein (Pederson et al. 1986). However, this modified phaseolin achieved less than 1% of the expression level of normal phaseolin in transformed seeds. Recently, it has been found that this insertion was made in part of a major structural element of the phaesolin trimer (Lawrence et al.

1990). Therefore, an inclusion of 15 residues at this site could distort the structure at the tertiary and/or quaternary level.

Lysine- and tryptophan-encoding oligonucleotides were introduced at several positions into a 19-kd a-type zein complementary DNA by oligo-nucleotide-mediated mutagenesis (Wallace et al. 1988). Messenger RNA for the modified zein was synthesized in vitro and injected into *Xenopus laevis* oocytes. The modified zein aggregated into structures similar to membrane-bound protein bodies. This experiment suggested the possibility of creating high-lysine corn by genetic engineering.

There are alternative approaches that might be more practical. One of these is to transfer heterologous storage protein genes that encode storage proteins with higher levels of the desired amino acids. For this purpose, a chimeric gene encoding a Brazil nut methionine-rich protein that contains 18% methionine has been transferred to tobacco and expressed in the developing seeds (Altenbach et al. 1989). The remarkably high level of accumulation of the methionine-rich protein in the seed of tobacco results in a significant increase in methionine levels of ~ 30%.

The maize 15-kd zein structural gene was placed under the regulation of French bean β-phaseolin gene flanking regions and expressed in tobacco (Hoffmann et al. 1987). Zein accumulation as high as 1.6% of the total seed protein was obtained. Zein was found in roots, hypocotyls, and cotyledons of the germinating transgenic tobacco seeds. Zein was deposited and accumulates in the vacuolar protein bodies of the tobacco embryo and endosperm. The storage proteins of legume seeds such as the common bean (*Phaseolus vulgaris*) and soybean (*Glycine max*) are deficient in sulfur-containing amino acids. The nutritional quality of soybean could be improved by introducing and expressing the gene encoding methionine-rich 15-kd zein (Pederson et al. 1986).

A 292-bp synthetic gene (HEAAE I = high essential amino acid encoding) that encoded a protein domain high in essential amino acid was expressed as a CAT-HEAAE I fusion protein in potato (Jaynes et al. 1986; Yang et al. 1989). However, structural instability limited the high-level expression of this fusion protein in the potato system. Also, the content of essential amino acids was diluted to less than 40% of the original encoded protein by constructing this fusion.

There are several precautions that should be considered in engineering storage proteins (Larkins 1983). First, in vitro mutational change must not be in regions of the protein that perturb the normal protein structure; otherwise, the proteins might be unstable. Second, when attempting to increase nutritional quality by introducing a gene encoding a heterologous protein in crop plants, it is important that the protein encoded by an introduced gene does not produce any adverse effects in humans or livestock, the ultimate consumers of the engineered seed proteins (Altenbach et al. 1989).

Finally, it is critical that the amino acids present in the introduced protein are able to be utilized by the animal for growth and development.

DE NOVO DESIGN OF PROTEINS

Recently, a new field in protein research, *de novo* design of proteins, has made remarkable progress due to a better understanding of the rules that govern protein folding and topology. Protein design has two components: the design of activity and the design of structure. This review concentrates on the design of structurally stable storage proteinlike proteins.

The usual approach for the design of helical bundle proteins consists of linking sequences with a propensity for forming an α-helix via short loop sequences to get linear polypeptide chains. This chain can fold into the predetermined ''globular-type'' tertiary structure in aqueous solution (Mutter 1988; Degrado, Wasserman, and Lear 1989). The α-helical secondary structures are stabilized by interatomic interactions that can be classified according to the distance between interacting atoms in the sequence of the protein (Degrado, Wasserman, and Lear 1989).

Short-range interactions account for different amino acids having different conformational preferences. Both statistical (Chou and Fasman 1978) and experimental (Sueki et al. 1984) methods show that residues such as glu, ala, and met tend to stablize helices, whereas residues such as gly and pro are destabilizing. However, these intrinsic preferences are not sufficient to determine the stability of helices in globular proteins.

Analysis of the free-energy requirements for helix initiation and propagation indicates that peptides of 10 to 20 residues should show little helix formation in water (Bierzynski, Kim, and Baldwin 1982) when the Zimm–Bragg equation (Zimm and Bragg 1959) is used, with parameters (s and S) determined by host–guest experiments where s is the helix nucleation constant, n is the number of H-bonded residues in the helix, and S is an average stability constant for one residue: $sSn - 1/(S-1)$.

Nevertheless, the 13 amino acid C-peptide obtained from RNase A does show measurable helicity ($\sim 25\%$) at low temperature (Bierzynski, Kim, and Baldwin 1982; Brown and Klee 1971). The stability of this peptide is 1000-fold greater than the value calculated from the Zimm–Bragg equation. Specific side-chain interactions, factors that are not considered in the Zimm–Bragg model, are responsible, at least in part, for the fact that the C-peptide is much more helical than predicted (Scherega 1985).

Medium-range interactions are responsible for the additional stabilization of secondary structures (Degrado, Wasserman, Lear 1989). Interaction between the side chains are regarded as important medium-range interactions (Shoemaker et al. 1987; Marqusee and Baldwin 1987). These include electrostatic interactions, hydrogen bonding, and the perpendicular stacking of aromatic residues (Blundell et al. 1986). An α-helix possesses a dipole moment as a result of the alignment of its peptide bonds. The positive and negative ends of the amide group dipole point toward the helix pair NH_2-terminus and COOH-terminus, respectively, giving rise to a significant macrodipole. Appropriately charged residues near the ends of the helix can favorably interact with the helical dipole and stabilize helix formation. It was estimated that the electrostatic interaction between a pair of antiparallel α-helices is about 20 kcal/mol less than a parallel α-helices pair (Hol and Sanders 1981). Hydrogen bonds between side chains and terminal helical $N-H$ and $C=O$ groups also participate in the stabilization of helical structure (Richardson and Richardson 1988; Presta and Rose 1988; Richardson and Richardson 1989).

Protein structures contain several long-range stabilizing interactions that include hydrophobic and packing interactions, and hydrogen bonds. Among these, the hydrophobic effect is a prime contributor to the folding and stabilizing of protein structures. The driving force for helix formation in RNase A arises from long-range interactions between C-peptide and S-protein, a large fragment of the protein from which C-peptide was excised (Komoriya and Chaiken 1982).

The role of hydrophobic interactions in determining secondary structures was studied for a series of peptides containing only glu and lys in their sequence (Degrado and Lear 1985). Glu and lys residues were chosen as charged residues for the solvent-accessible exterior of the protein to help stabilize helix formation by electrostatic interaction.

STABILITY OF DESIGNED PROTEINS

Hydrophobic residues often repeat every three to four residues in an α-helix and form an amphiphilic structure (Degrado, Wasserman, and Lear 1989). Amphiphilicity is important for the stabilization of the secondary structures of peptides and proteins that bind in aqueous solution to extrinsic apolar surfaces, including phospholipid membranes, air, and the hydrophobic binding sites of regulatory proteins (Degrado and Lear 1985). This amphiphilic secondary structure can be stabilized relative to other conformations by self-association. Therefore, short peptides often form the α-helix in water only because the helix is amphiphilic and is stabilized by peptide aggregation along the hydrophobic surface. Natural globular proteins are folded by a similar mechanism, involving hydrophobic interaction between neighboring segments of secondary structure (Presnell and Cohen 1989). Using the concept of an amphiphilic helix, Degrado and coworkers have successfully built peptide–hormone analogs with minimal homology to the native sequences. These peptides, like the native ones, are not helical in solution, but do form helices at the hydrophobic surfaces of membranes.

Designed synthetic peptides have been used to show how hydrophobic periodicity in a protein sequence stabilizes the formation of simple secondary structures such as an amphiphilic α-helix (Ho and Degrado 1987). The strategies used in the design of the helices in the four-helix bundles are that the helices should be composed of strong helix-forming amino acids, and the helices should be amphiphilic; that is, they should have an apolar face to interact with neighboring helices and a polar face to maintain water solubility of the ensuing aggregates. The results show that hydrophobic periodicity can determine the structure of a peptide. Therefore, the peptides tend to have random conformations in very dilute solution, but form other secondary structures when they self-associate (at high concentration) or bind to the air–water surface.

The free energy associated with dimerization or tetramerization of the designed peptides could be experimentally determined from the concentration dependence of the CD spectra for the peptides (Degrado, Wesserman, and Lear 1989; Lear, Wasserman, and Degrado 1988; Degrado and Lear 1985). At low concentrations, the peptides were found to be monomeric and have low helical contents, whereas at high concentration they could self-associate and stabilize the secondary structure. Therefore, possible hairpin loops between helices can affect the stability of the secondary structure by enhancing the self-association between the helical monomers. A strong helix breaker (Chou and Fasman 1978; Kabsch and Sander 1983; Sueki et al. 1984; Scheraga 1978) was included as the first and last residue to set the stage for adding a hairpin loop between the helices. A single proline residue appeared capable of serving as a suitable link if the C and N terminal glycine residues are slightly unwound. Glycine lacks a β-carbon, which is essential for the reverse turn where positive dihedral angles are required. The pyrrolidine ring of proline constrains its f dihedral angle $-60°$. Thus, proline is destabilizing at positions where significantly different backbone torsion angles are required. This amino acid, as well as glycine, has a high tendency to break helices and occurs frequently at turns (Creighton 1984).

The direct evidence for stabilization of protein structure by adding the linking sequence was observed by comparing the guanidine denaturation curve for a monomer, dimer, and tetramer (Degrado, Wasserman, and Lear 1989). The gene-encoding tetrameric protein was expressed in *E. coli* and purified to homogeneity. In the series of mono-, di-, and tetramer, the stability toward guanidine denaturation increases concomitantly with the increase in covalent cross-links between helical monomer. At equivalent peptide concentrations, the midpoints of the denaturation curves occurred at 0.55, 4.5, and 6.5 M guanidine for the mono-, di, and tetramer. Furthermore, as the number of covalent cross-links was increased, the curves became increasingly cooperative. Thus, the linker sequence stabilized the formation of the four helix structures at low concentration of the peptides (< 1 mg/ml).

Structural stability of proteins is directly related to in vivo proteolysis (Parasell and Sauer 1989). Proteolysis depends on the accessibility of the scissile peptide bonds to the attacking protease. The sites of proteolytic processing are generally in relatively flexible interdomain segments or on the surface of the loops, in contrast to the less accessible interdomain peptide bonds (Neurath 1989). This suggests that the stability of the folded state of the protein is the most important determinant for its proteolytic degradation rate. The effect of a folded structure on the proteolytic degradation has been proved by several experiments. First, proteins that contain amino acid analogs or are prematurely terminated are often degraded rapidly in the cells (Goldberg and St. John 1976). Second, there are good correlations between the thermal stabilities of specific mutant proteins and their rates of degradation in *E. coli* (Pakula and Sauer 1986; Parasell and Sauer 1989). Finally, second-site suppressor mutations that increase the thermodynamic stability of unstable mutant proteins have also been shown to increase resistance to intracellular proteolysis (Pakula and Sauer 1989). The solubility of proteins could also affect their proteolytic resistance, since some proteins aggregate to form inclusion bodies that escape proteolytic attack (Kane and Hartley 1988).

Metabolic stability is another factor influencing the in vivo stability of proteins. Usually, damaged and abnormal proteins are metabolically unstable in vivo (Finley and Varshavsky 1985; Pontremoli and Melloni 1986). In eukaryotes, covalent conjugation of ubiquitin with proteins is essential for the selective degradation of short-lived proteins (Finley and Varshavsky 1985). It was found that the amino acid at the amino terminus of the protein determined the rate of ubiquitination (Bachmair, Finley, and Varshavsky 1986). Both prokaryotic and eukaryotic long-lived proteins have stabilizing amino acids such as methionine, serine, alanine, glycine, threonine, and valine at the amino terminus end. On the other hand, amino acids such as leucine, phenylalanine, aspartic acid, lysine, and arginine destabilize the target proteins.

DESIGN OF THE ARTIFICIAL STORAGE PROTEIN (ASP 1)

We designed the synthetic protein ASP 1 to contain a high content of those amino acids that are essential to the diet of humans and animals. The optimized content of essential amino acids for this new protein was obtained empirically by obtaining the amounts of essential amino acids necessary for normal metabolism of the human being, from infants to the aged (Table 1-1). We also determined the "deficiency values" or the ratios of deficient essential amino acids (EAA) for the 10 primary crops people consume throughout the world (Table 1-2). From these data, we then found the ratio of essential amino acids needed to totally complement each particular plant foodstuff. We merely averaged these values and came up with a set of numbers we call the "average ratio for all crops idealized to the ASP 1 monomer" (Fig. 1-3). This set of numbers represents the ratio of essential amino acids necessary to complement the deficiencies found in all 10 crops for all human age groups.

The structural stability of a protein is important in determining its susceptibility to proteolysis. Most native proteins are relatively resistant to cleavage by proteolytic enzymes, whereas denatured proteins are much more sensitive (Pace and Barret 1984). Several findings suggest that the stability of a folded protein is an important determinant of its rate of degradation. Therefore, in addition to improved nutritional quality, ASP 1 has been designed to have a stable storage proteinlike structure in plants. Its design is based on the structurally well-studied corn storage zein proteins (Z19 and Z22), which are comprised of nine repeated helical units (Agros et al. 1982). Each helical unit, 16 to 26 amino acids long, of zein is flanked by turn regions and forms an antiparallel helical bundle. Most of the amino acids in the helices are hydrophobic residues. On the other hand, ASP 1 is comprised of four helical repeating units, each 20 amino acids long (Fig. 1-4). Increased gene copy number by concatenation can increase the protein yields. At the same time, gene concatenation gives the increased molecular

TABLE 1-1   Essential Amino Acid Requirements in Grams per day of Children[a] and Adults[b]

	Infant	Child	Child	Average Child	Adult	Adult	Average Adult	Overall Average
Ile	0.258	0.524	0.879	0.554	0.754	0.754	0.754	0.654
Leu	0.521	1.234	1.382	1.046	1.102	1.102	1.102	1.074
Lys	0.370	1.150	1.382	0.967	1.020	1.020	1.020	0.994
Met + Cys	0.235	0.468	0.691	0.465	0.986	0.986	0.986	0.725
Phe + Tyr	0.403	1.178	1.124	0.902	1.102	1.102	1.102	1.002
Thr	0.241	0.636	0.879	0.585	0.522	0.522	0.522	0.554
Trp	0.095	0.185	0.240	0.173	0.260	0.260	0.260	0.217
Val	0.308	0.655	0.785	0.583	0.754	0.754	0.754	0.668

[a]Ages 3 months, 5 years, 10 years.
[b]Ages 25 years and 75 years.

TABLE 1-2 Essential Amino Acid Deficiency Ratios of the Ten Major Crop Plants Consumed by Humans

	Wheat	Corn	Rice	Barley	Sorghum	Cassava	Taro	Sweet Potato	Potato	Plantain
Ile	1.72	2.23	1.71	1.94	1.51	1.88	1.69	1.92	1.68	1.33
Leu	1.85	1.25	1.57	1.98	0.85	2.13	1.64	2.70	2.45	2.10
Lys	4.08	3.36	3.03	3.68	4.54	2.83	2.29	2.96	2.08	2.24
Met + Cys	1.73	2.41	3.86	2.53	2.55	3.18	3.52	2.84	3.53	3.74
Phe + Tyr	1.20	1.30	1.10	1.32	1.49	1.58	2.37	1.29	1.67	1.71
Thr	2.14	1.67	1.75	2.01	1.89	1.58	1.55	1.62	1.54	2.28
Trp	1.61	2.20	1.78	1.37	1.69	1.02	1.41	1.37	1.60	1.39
Val	1.68	1.39	1.20	1.18	1.50	1.23	1.31	1.19	1.01	1.36

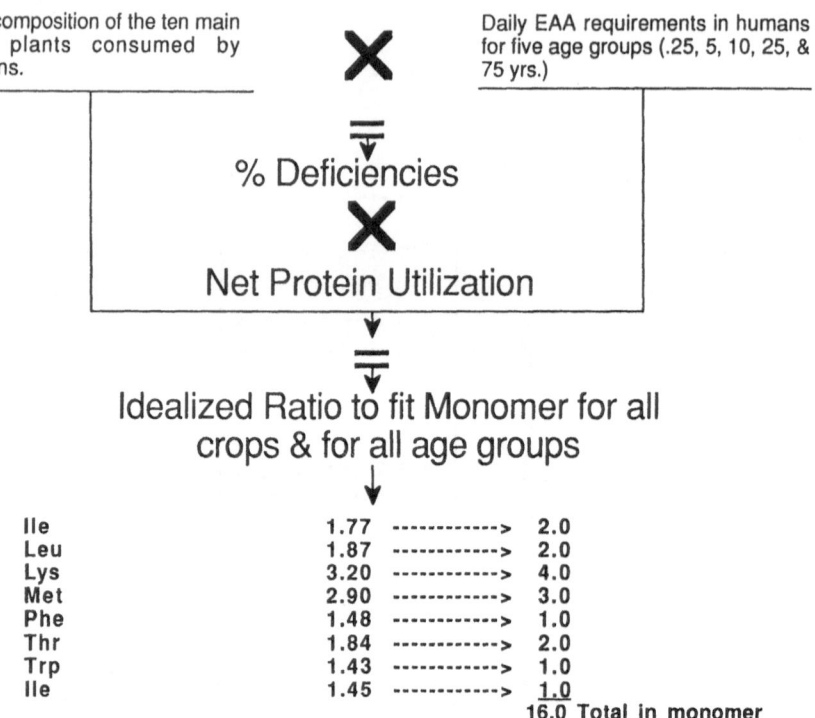

Ile	1.77 ------------>	2.0
Leu	1.87 ------------>	2.0
Lys	3.20 ------------>	4.0
Met	2.90 ------------>	3.0
Phe	1.48 ------------>	1.0
Thr	1.84 ------------>	2.0
Trp	1.43 ------------>	1.0
Ile	1.45 ------------>	1.0

16.0 Total in monomer

FIGURE 1-3.   How the composition ration of the ASP 1 monomer was calculated. The ratio is an overall average that takes into consideration all daily requirements for five age groups; the deficiencies of the ten major crop plants people consume (rice, wheat, barley, sorghum, maize, potato, sweet potato, plantain, cassava, and taro); and how well these plant foodstuffs are utilized as a protein source.

mass of the encoded protein. Such an increase in size and concatenation can significantly stabilize an otherwise unstable product (Shen 1984).

From the preceding set of numbers, we designed the ASP 1 protein. It has 1.8 times more of the essential amino acids compared to zein or phaseolin. The difference in MLEAA is much higher, containing three times more than phaseolin and 6.5 times more than zein (Table 1-3). The helical region of ASP 1 is amphipathic (hydrophobic residues clustered on one face of the helix, while hydrophilic residues are found on the other face) and is stabilized by several Glu–Lys salt bridges (Fig. 1-5). The helix breaker Gly-Pro-Gly-Arg has been used as a turn sequence. The design results in an antiparallel tetramer that achieves an extraordinarily stable secondary and tertiary structure even at low concentration.

The 284-bp gene (Fig. 1-6) encoding this novel peptide was chemically synthesized and cloned into an *E. coli* expression vector. This gene contains

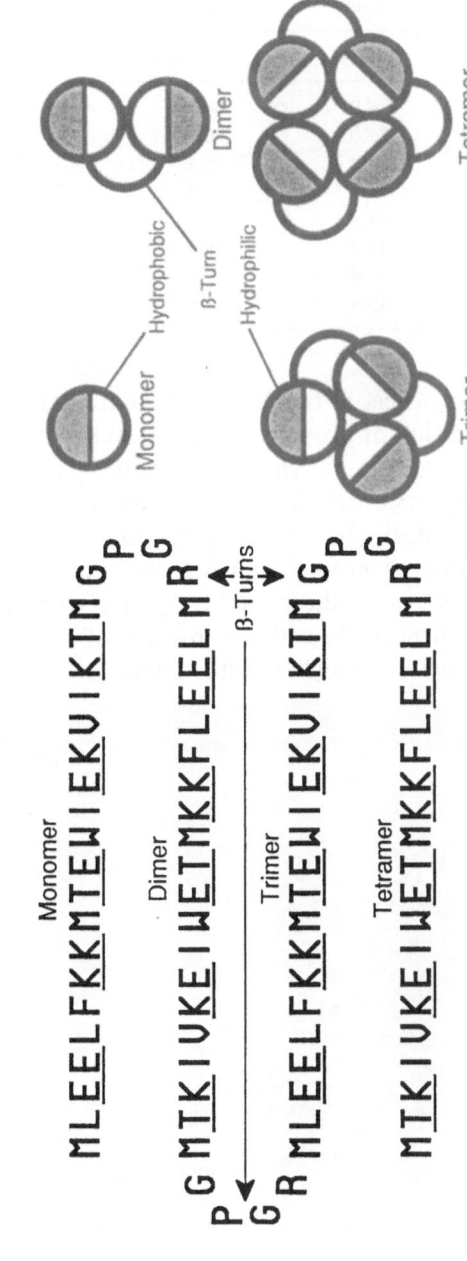

FIGURE 1–4. The sequence (*A*) and conformation (*B*) of the ASP 1 monomer to tetramer. In (*B*) the white regions represent the hydrophobic face, while shaded regions depict the hydrophilic face.

TABLE 1-3   Percentage of Essential Amino Acids (EAA)
and Percent of Most Limiting Essential Amino Acids (MLEAA)
of the ASP 1 Monomer to Tetramer when Compared
with Natural Foodstuffs

Protein	%EAA	%MLEAA	Molecular Weight
ASP 1	78.9	60.2	2,526
ASP 2	73.8	56.3	5,402
ASP 3	72.2	55.1	8,278
ASP 4	71.4	54.5	11,154
Egg	43.9	21.7	42,933
Milk	47.7	28.7	14,011
Bean	39.4	17.6	46,502
Corn	39.9	8.5	23,322

plant consensus sequences at the 5' end of the translation initiation site to
optimize the expression of proteins in vivo. It was placed under the control
of the 35S cauliflower mosaic virus (CaMV) promoter in order to permit
the constitutive expression of this gene in tobacco.

So far, few laboratories have succeeded in their attempts to obtain stable
expression of entirely *de novo* designed proteins in an *E. coli* system, and
none have been able to do the same in higher organisms. This chapter is
the first announcement of the stable expression of a *de novo* designed protein (ASP 1–tetramer) in a higher plant system.

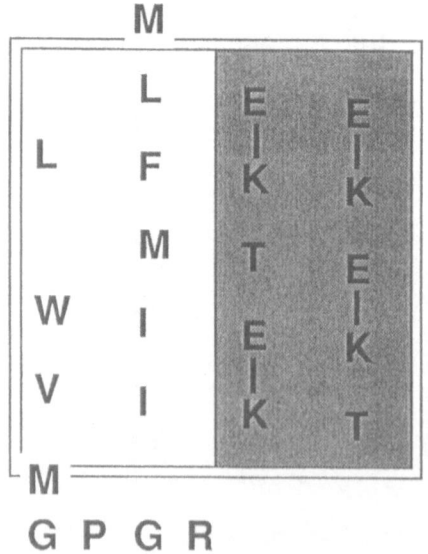

FIGURE 1-5.   A depiction of the amphiphilicity of the ASP 1 monomer where hydrophobic amino acids are in the white rectangle and hydrophilic in the shaded
rectangle. Note the interaction between the
Glu (E) and Lys (K) residues.

```
                START
BamHI        ┌─────►
GATCC AACAA TGCTTGAAGAGCTGTTCAAAAAGATGACCGAGTGGATCGAGAAAGTGATCAAAA
CTAGG TTGTT ACGAACTTCTCGACAAGTTTTTCTACTGGCTCACCTAGCTCTTTCACTAGTTTT
      KOZAK

                    XhoI
CGATGGGACCAGGCAGGATGCTCGAGGAGCTGTTCAAAAAGATGACCGAGTGGATCGAGAAAGT
GCTACCCTGGTCCGTCCTACGAGCTCCTCGACAAGTTTTTCTACTGGCTCACCTAGCTCTTTCA


                    XhoI
GATCAAAACGATGGGACCAGGCAGGATGCTCGAGGAGCTGTTCAAAAAGATGACCGAGTGGATC
CTAGTTTTGCTACCCTGGTCCGTCCTACGAGCTCCTCGACAAGTTTTTCTACTGGCTCACCTAG


                          XhoI    SacI   DraI
GAAAGTGATCAAAACGATGGGACCAGGCAGGATGCTCGAGGAGCTCTTTAAAAAAATGACTGAG
CTTTCACTAGTTTTGCTACCCTGGTCCGTCCTACGAGCTCCTCGAGAAATTTTTTTACTGACTC


                   STOP
TGGATCGAAAAAGTGATCAAAACTATGTAGGAATT
ACCTAGCTTTTTCACTAGTTTTGATACATCCTTAA
                   EcoRI
```

FIGURE 1-6.   The DNA sequence of the ASP 1 tetramer form of the gene showing restriction sites and translational start and stop points of the gene. The shaded rectangle is a "Kozak" sequence that has been shown to enhance translation.

PREDICTION OF THE STRUCTURE OF ASP 1

The secondary structures of the ASP 1 monomer and tetramer were predicted by PREDICT-SECONDARY in β-SYBYL. The precentage of α-helix content predicted by information theory showed a higher α-helix content compared to the other two prediction methods (Bayes statistic and neural net) in PREDICT-SECONDARY. The predicted secondary structures by information theory gave 100% helical content for the monomer and 74% for the tetramer (Table 1-4).

TABLE 1-4   Prediction of the Secondary Structure of the ASP 1 Monomer/Tetramer

Methods	α-Helix	β-Sheet	Coil
Neural net	80/70	0.0	20/30
Information theory	100/74	0.0	0/26
Bayes statistic	55/68	0.0	45/32

However, the accuracy of the three widely used prediction methods ranged from 49% to 56% for prediction of three states; helix, sheet, and coil (Kabsch and Sander 1983). This inaccuracy might be due to the small size of the database and/or the fact that secondary structure is determined by tertiary interactions that are not included in the local sequences. For further predictions of structure, the structures predicted by information theory were energy minimized using SYBYL MAXIMIN2.

A perfect amphiphilic α-helical conformation was predicted for the ASP 1-monomer after minimization. The tertiary structure of the ASP 1-tetramer after minimization showed the antiparallel conformation as was designed. These minimization results suggested the high probability of stable secondary structure (α-helix and β-turn) formation of the ASP 1 monomer and tetramer. However, it might be too early to conclude the tertiary structure of the ASP 1 tetramer with minimization only, without considering longer range interactions, which are the most important determinant of protein folding.

The predominant driving force for folding the ASP 1 tetramer might be the longer range hydrophobic interactions between the α-helical monomers, because ASP 1 was designed to be amphiphilic. However, no currently available force field for the minimization of tertiary structure contains these parameters and could not give the perfect predicted tertiary structure of the protein.

Actually, minimization schemes alone have failed to predict chain-folding accurately (Fasman 1989). Therefore, we might be able to obtain indications of much more stable secondary and tertiary structures if we considered this factor for the minimization of ASP 1 tetramer.

STRUCTURAL ANALYSIS OF ASP 1 PROTEIN

The structural stability of ASP 1-monomer and tetramer could not be determined by minimization only. Therefore, the stability of the α-helical secondary structure of ASP 1 monomer was investigated. HPLC analysis of the gel-filtered synthetic ASP 1 monomer showed that purity was more than 90% and amino acid analysis of the purified fraction gave the expected molar ratios. This fraction was also analyzed by mass spectrometry, and the molecular weight peak corresponding to the ASP 1 monomer (2896.5) was present (Fig. 1–7). Since the structural stability of ASP 1 monomer and tetramer could not be determined by minimization only, the stability of the α-helical secondary structure of ASP 1 monomer was investigated by circular dichroism (CD) analysis. CD spectra of ASP 1 monomer showed the typical pattern of α-helical proteins with double minima at 208 and 222 nm in aqueous solution (data not shown). The stability of the secondary struc-

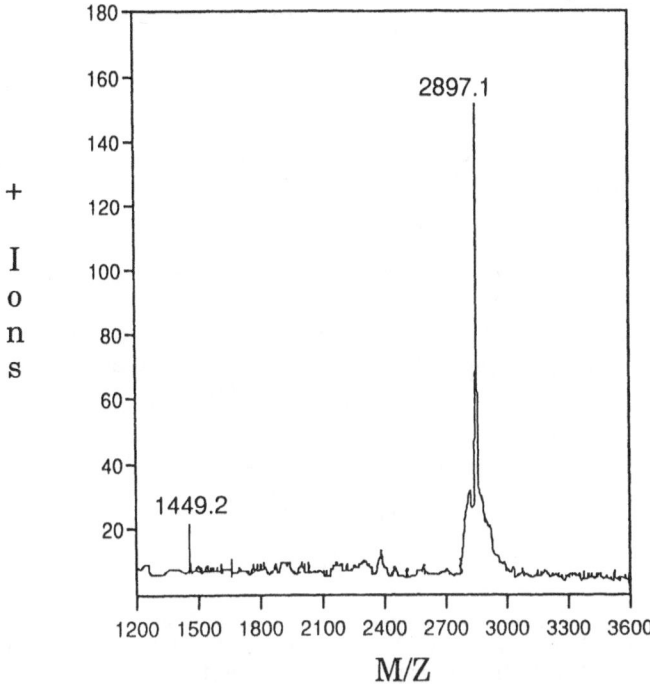

FIGURE 1-7. Mass spectrum of ASP 1 monomer indicates a clear 2897.1 molecular ion (MI) peak that is precisely the molecular weight expected. Oftentimes, an MI/2 is seen in spectra, and 1449.2 was detected.

ture can be induced by the inter-molecular interaction between the helical chains (Degrado, Wasserman, and Lear 1989). Therefore, stable aggregation between monomers, presumably through hydrophobic interactions, could stabilize the helical structure. Besides, proper packing of the apolar side chains and proper electrostatic interaction might play important roles in stabilizing the secondary structure of ASP 1. The stable interaction among the monomeric ASP 1 molecules is an important determinant for the proper folding into the tertiary structure of the ASP 1 tetramer. Therefore, the self-association capability of the ASP 1 monomers was investigated by using size-exclusion chromatography. The hydrodynamic behavior of this peptide showed that it was aggregated into a hexamer form with an apparent molecular weight of about 17 kd (Fig. 1-8). This hexameric aggregate could be maintained in either low or high ionic strength solutions. This result provides proof of the stable globular-type tertiary structure formation of tetrameric ASP 1.

Three potential β-turn (Gly-Pro–Gly-Arg) sequences were inserted between four monomers for the ASP 1-tetramer construction. The β-turn

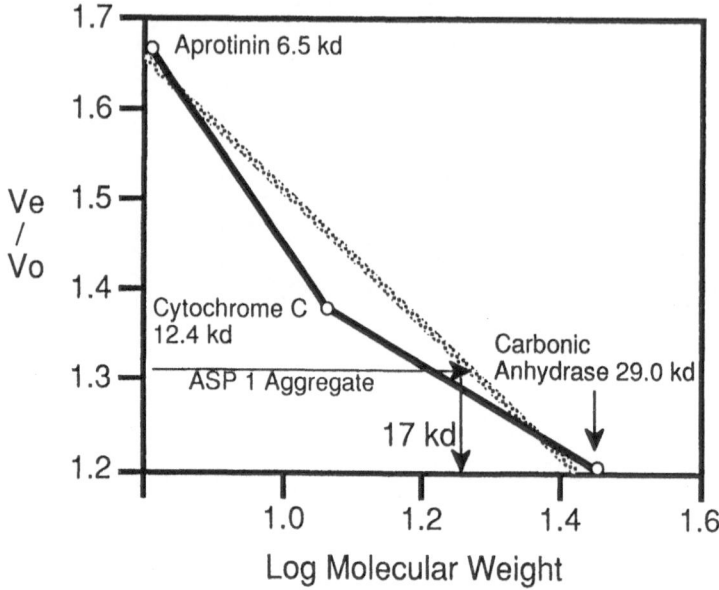

FIGURE 1-8. Calibration curve for the 1.6-cm × 90-cm G50F Sephadex column used in the size-exclusion chromatography of the aggregation of the ASP 1 monomer. The apparent molecular weight (17 kd) of the aggregated form of ASP 1 monomer was determined by this standard curve.

could play an important role for structural stability of the ASP 1 tegramer when it is expressed in vivo. It can also help stabilize tertiary structure formation. The interactions between the helical monomers might be much faster due to the proximate effect when they are connected. This proximate effect might be critical for folding at the low concentrations of ASP 1 tetramer that are possible when they are expressed in vivo. At the same time, the stability of the secondary structure is increased by the hydrophobic interactions between helical monomers. In addition, this β-turn sequence has a tryptic digestion site (Gly-Arg) that could increase the digestibility of this protein when it is consumed by animals.

The stability of the folded structure of a protein has a close relation to its proteolytic degradation rate (Pace and Barret 1984; Pakula and Sauer 1986; Parasell and Sauer 1989; Pakula and Sauer 1989). In this respect, we expected high stability of folded ASP 1 tetramer against proteolytic degradation when it is expressed in vivo. Stable quaternary structure is essential for the formation of protein bodies of storage proteins in zein or phaseolin (Lawrence et al. 1990). These higher order structures can be achieved through the interaction and close packing of the stable tertiary structures. The major driving force for this quaternary structure formation is also hydrophobic interaction between the tertiary structures. At this moment, designing and predicting of the quaternary structure is not easy, but our data suggest that precisely this might be occurring with ASP 1 tetramer.

INTRODUCTION OF ASP 1 GENE
INTO TOBACCO

The correct insertion and orientation of the pBI derivative containing the ASP 1 tertramer, was screened by EcoRI and HindIII digestion (it was found in *E. coli* that the most stable form of the gene was the tetramer form). The EcoRI digestion gave a fragment of the expected size, 3.2 kb, which consisted of 3'NOS of ASP 1 and the GUS gene (data not shown). Also, the ASP 1 gene with its 35S promoter and 3'NOS sequences was detected as a 1.4-kb band by HindIII digestion. Stable transformation of the ASP 1 gene into *A. tumefaciens* LBA4404 was confirmed by HindIII digestion of isolated plasmid DNA. It could be isolated from agrobacterium and detected by enzyme digestion because pBI121 is a binary vector. Leaf discs, transformed with LBA4404 carrying the ASP 1 gene, gave about 5 to 7 shoots two to three weeks after infection. A total of 565 kanamycin-resistant shoots were regenerated from 120 leaf discs. These shoots were excised from the leaf discs and transferred to new media to grow several more weeks, and then transferred to rooting media. After three weeks in the rooting medium, 126 rooted shoots were analyzed for β-glucuronidase (GUS). Root tips of 56 out of 126 plants showed various levels of GUS activity. Not all the kanamycin-resistant shoots showed the GUS positive result. Although kanamycin resistance was due to the expression of neomycin phosphotransferase (NTP II gene), regeneration of nontransgenic shoots in the presence of kanamycin was seen. Therefore, escapes from the screening based on kanamycin sensitivity might have occurred in the nontransformed plants, making them kanamycin resistant.

Thirty-six plantlets that showed high levels of β-glucuronidase activity were transplanted into jiffy pots. After establishment of the plants, a more accurate fluorogenic assay for GUS activity was done to quantify the expression level of this gene (Table 1–5). Some of these transformed tobacco plants showed higher levels of β-glucuronidase activity compared to other plants. The level of expression might be primarily affected by whether the gene is incorporated into an active or inactive site of chromatin. Activity of chromatin, methylation of DNA, and nuclease hypersensitivity are closely related to each other. It has been found that the nuclease hypersensitive sites correlate to active transcription (Gross and Garrard 1987). The degree of methylation of DNA is inversely related to gene expression. Furthermore, if the gene is located near the plant's endogenous promoter or en-

TABLE 1-5   The GUS Activities of Transformed Seedlings of Both Control (pBI121 #2 & Wildtype #37) and Various ASP 1 Tetramer Plants

Transgenic Plant	Gus Activity
ASP 1 #1	200
ASP 1 #9	315
ASP 1 #11	3,790
ASP 1 #13	360
ASP 1 #17	2,400
ASP 1 #29	200
pBI 121 #2	320
Wildtype #37	10

Note: Gus activity was measured as pmole 4-methyl umbelliferone produced per milligrams protein per minute with an excess of 4-methyl umbelliferone glucuronide.

hancer sites, the level of expression of this gene will be increased by these nearby enhancing factors. Therefore, the difference in the levels of GUS activity between the transformed plants might be due to this positional effect, which was determined by the sites of incorporation of this gene into the tobacco genome.

ANALYSIS OF TRANSFORMED PLANTS

DNA Analysis

Although GUS activity and kanamycin resistance are good indicators of transformation, rearrangement of the T-DNA after incorporation in the plant genome can inactivate or silence the other genes transferred. Therefore, correct incorporation of the ASP 1 gene in the tobacco genome was determined by southern blotting. A distinct 1.4 kb-HindIII band appeared in seven samples out of nine tobacco genomic DNA samples analyzed, but did not appear in negative control samples, using the ASP 1 tetrameric fragment as a probe (data not shown). As a positive control, and to check the copy number, HinIII-digested plasmid pBIASP 1 tetramer was also loaded, corresponding to 1 and 5 copy number of the inserted gene in tobacco DNA. Multiple positive bands were observed together with the expected size band of 1.4 kb, from most of the transformed plants. All the extra bands that appeared were bigger than 1.4 kb and showed different patterns between the individual plants. These results suggested that the ASP 1 gene, alone or with neighboring genes, might be inserted into several sites in the chromosomes with or without rearrangement. The copy number of the correct band varied among the plants, and ranged from 1 to 5 by densitometeric measurement. The copy number of a gene can affect its expression level. The impact of copy number on the extent of expression varies from one system to another. In some cases there are positive correlations to their expression (Scott and Draper 1987; Stockhaus et al. 1987), but not always (Jones and Gilbert 1987; Sanders et al. 1987). Therefore, it should not be expected that all the copies of the gene are equally active because the position of each copy of the gene incorporated into the genome can also affect the level of the transcription. However, as the number of chromosomal sites containing foreign DNA increases, the likelihood that any one of the pieces of DNA would integrate into a transcriptionally active region is increased.

Kanamaycin Gene Segregation Test

The first-generation progency from the self-fertilized transformed parents were tested for kanamycin gene segregation. Since the integrated T-DNA is passed on as a dominat Mendelian trait, copy number of the ASP 1 gene can be determined by the kanamycin segregation pattern of the progenies. The results (Table 1–6) showed that most transformed plants have multicopies of the ASP 1 gene. The progenies of the transformed plants carrying a single, double, or triple genetic NPTII loci are expected to segregate in 3:1, 15:1, or 63:1 ratios. Therefore, #2 and #13 plants have one NPTII loci, #1, #11, and #29 plants have two NPTII loci, and #17 plant has more than three loci of kanamycin gene in the chromosome.

RNA Analysis

Efficient transcription of inserted ASP 1 genes in tobacco plants was tested by northern analysis. The polyA RNA was analyzed using the ASP 1 tetra-

TABLE 1-6   Kanamycin Gene Segregation of Control (pBI121 #2) and Various ASP 1 Tetramer Seedlings

Transgenic Plant	Kn(r)	Kn(s)	Kn(r)/Kn(s)
ASP 1 #1	136	8	15:1
ASP 1 #11	127	8	15:1
ASP 1 #13	112	37	3:1
ASP 1 #17	175	1	175:1
ASP 1 #29	131	9	15:1
pBI 121 #2	107	34	3:1

Note: Single genetic loci are found in the control pBI121 #2 and ASP 1 #13, double genetic loci are found in ASP 1 #'s 1, 11, and 29. More than three genetic loci can be found in ASP 1 #17.

mer probe. The correct gene size transcribed was about 490 bases, which consisted of 30 bases upstream and 170 bases downstream of the ASP 1 tetramer gene. In addition to this message, eukaryotic mRNA contains different sizes of polyA. Therefore, the expected size of the ASP 1 tetramer message should be around 600 + ~100 bases long. Bands were observed that correspond to this expected size from all the samples that were analyzed (data not shown). However, the levels of transcription of the ASP 1 genes were dramatically different among the transformed plants. Transformed plant #17 accumulated 5- to 50-fold more transcripts compared to the other transformed plants. Such differences in accumulation could be explained by the effect of position or multicopy insertion. The expression levels of the ASP 1 gene and its neighboring GUS gene correlated with each other in some transformed plants (#17), but not in all. These results suggested that the level of expression can be dramatically different between the two closely connected genes. Multiple transcripts with different sized bands (500–700 bases) were observed from several transformed plants. This result might be due to multiple insertion of the ASP 1 gene into the tobacco genome. These inserted genes might be rearranged but still produce the transcripts. Another possibility might be strong secondary structure that could be formed due to the four directly repeated sequences of the tetrameric ASP 1 transcripts, which could cause the different mobilities dependent on the secondary structure.

Expression of ASP 1

Polyclonal antibody raised against synthetic ASP 1 monomer was used to detect the production of stable ASP 1 protein in tobacco. High levels of the tetrameric form (11.2 kd) of the ASP 1 protein were detected from plant #17 by western blot analysis (data not shown). Therefore, direct correlation was found between gene copy number, number of genetic NPTII loci, GUS expression, accumulation of ASP 1 transcript, and protein expression level in the case of plant #17. Some heterologous seed proteins undergo specific degradation when expressed in transgenic plants. A significant amount of the immunoreactive protein accumulated in tobacco seed expressing the phaseolin gene is smaller than the final processed protein (Sengupta et al. 1985). A similar result was found when β-conglycinin was expressed in transgenic petunia (Beachy et al. 1985). In contrast to these results, the ASP 1 protein appears to be quite stable in transgenic tobacco plants.

Amino acid and total protein analyses were conducted on leaf tissue from several of the transgenic plants that produced detectable levels of ASP 1. Surprisingly, we found that the overall levels of all amino acids were increased with some of the plants being remarkably high (Fig. 1-9). This

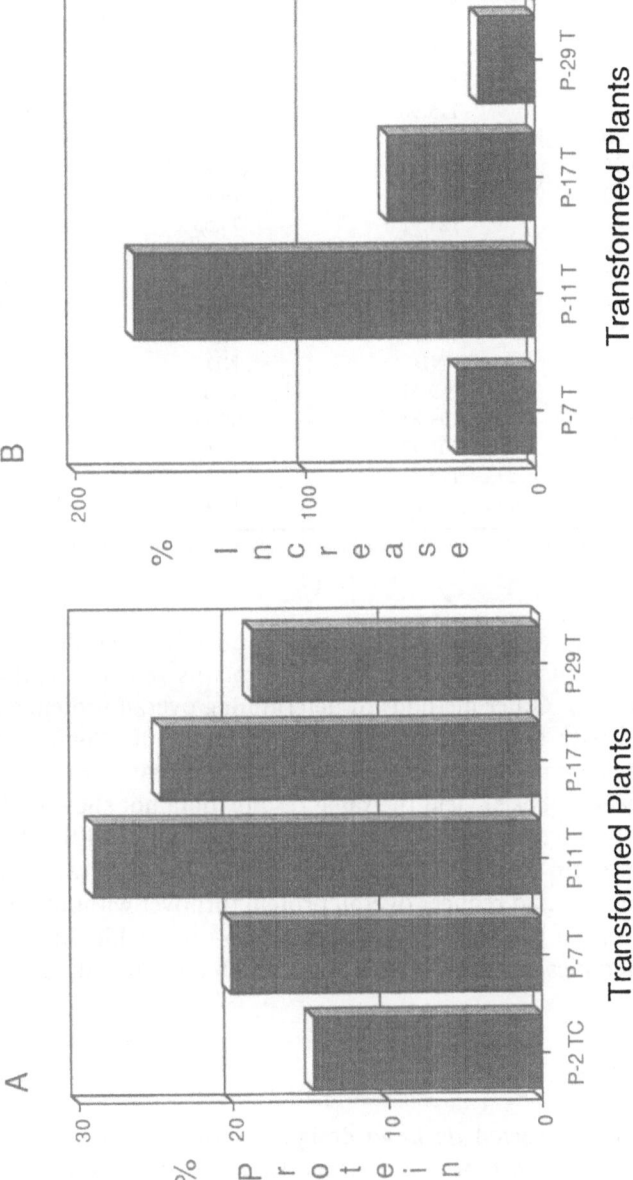

FIGURE 1-9. Overall protein content determined by amino acid analysis from control (P-2 TC) and various ASP 1 tetramer plants. These data were derived from seedlings obtained from transformed mother plants. A minimum of four separate assays were used and the variation was no more than 30%.

TABLE 1-7   Percentage of the Amino Acids above the Control (pBI121 #2) from Protein Isolated from Various ASP 1 Tetramer Seedlings when Compared to the Control

	% of 7 above C	% of 11 above C	% of 17 above C	% of 29 above C
Asx	60.55	80.59	47.00	35.00
Glx	65.26	56.18	46.42	20.68
Ser	30.00	109.67	73	28.00
Gly	14.46	115.96	78.01	23.69
His	31.30	94.27	63.74	27.23
Arg	39.06	86.5	58.28	23.24
Thr	31.00	106.55	79.48	23.44
Ala	6.21	123.91	76.55	27.54
Pro	11.68	114.53	73.65	19.66
Tyr	254.95	261.26	236.49	121.02
Val	14.45	80.06	53.32	23.70
Met	ND	ND	ND	ND
Cys	ND	ND	ND	ND
Ile	19.86	69.51	45.3	22.65
Leu	3.95	99.81	68.17	22.54
Phe	2.65	101.77	72.42	14.45
Lys	−25.90	119.51	79.34	3.61
Trp	ND	ND	ND	ND

Note: These numbers where derived from amino acid analysis. ND refers to not determined.

rather disconcerting result has been repeated numerous times and the overall levels of all amino acids in the transgenic plants remain significantly elevated (Table 1-7). Other methods of determining overall protein content have been used with similar trends observed (Fig. 1-10). Comparison of total protein densitometric values derived from SDS-PAGE of equivalent samples (on a weight basis) yield the same results (data not shown). At this time, we can offer no definitive explanation. Therefore, in addition to being a very stable protein in a plant cell, ASP 1 must function as a general "protein-stabilizer," and reduces overall protein turnover without apparent deleterious effects to the plants, since there is no observable difference in growth characteristics in the plants producing high amounts of ASP 1 over control plants.

CONCLUSIONS

ASP 1 is the first reported *de novo* designed protein expressed in plant systems. We have found that it is possible to construct a protein of high nutritional value mimicking the well-known physical characteristics of plant storage proteins. Plans for future studies of tetrameric ASP 1 begin with further analysis of its structural properties such as: (1) stability against pro-

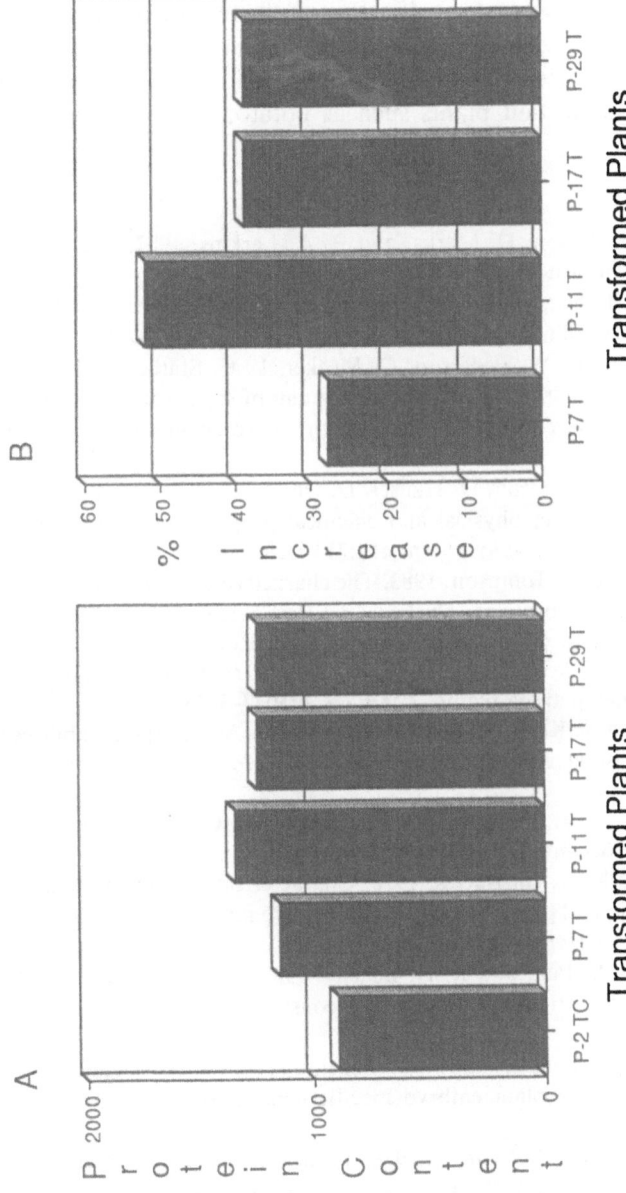

FIGURE 1-10. Overall protein content derived from biochemical measurements from equivalent samples (by weight) taken from leaves and control (P-2 TC) and various ASP 1 tetramer seedlings.

teolytic attack, and (2) solubility and aggregation pattern. Additionally we are intensively studying the import of its apparent ability to reduce overall protein turnover. After optimization of expression using a baculovirus expression system, large-scale purification of the tetrameric protein will allow small animal feeding studies to assess its biological value. Subsequent plans are to pursue high-level tissue-specific expression of this gene in the more economically important plants such as potato, soybean, and selected cereals.

References

Agros, P., K. Pederson, D. Marks, and B. A. Larkins. 1982. A structural model for maize zein proteins. *J. Biol. Chem.* **257:**9984–9990.

Agros, P., S. V. L. Naravana, and N. C. Nielsen. 1985. Structural similarity between legumin and vicillin storage proteins from legumes. *EMBO J.* **4:**1111–1117.

Altenbach, S. B., K. W. Pederson, G. Meeker, L. C. Staraci, and S. S. M. Sun. 1989. Enhancement of the methionine content of seed proteins by the expression of a chimeric gene encoding a methionine-rich protein in transgenic plants. *Plant Mol. Biol.* **13:**513–522.

Badley, R. A., D. Atkinson, H. Hauser, D. Oldani, J. P. Green, and J. M. Stubbs. 1975. The structure, physical and chemical properties of the soybean protein glycinin. *Biochim. Biophys. Acta* **412:**214–228.

Bartels, D., and R. D. Tompson. 1983. The characterization of cDNA clones coding for wheat storage proteins. *Nucleic Acid Res.* **11:**2961–2977.

Beachy, R. N., Z. L. Chen, R. B. Horsch, S. G. Rogers, N. J. Hoffman, and R. T. Fraley. 1985. Accumulation and assembly of soybean β-conglycinin in seeds of transformed petunia plants. *EMBO J.* **4:**3047–3053.

Bierzynski, A., P. S. Kim, and R. L. Baldwin. 1982. A salt bridge stabilizes the helix formed by isolated c-peptide of RNAse A. *Proc. Natl. Acad. Sci. (USA)* **79:**2470–2474.

Blundell, T. L., S. J. Thornton, S. K. Burley, and G. A. Petsco. 1986. Atomic interactions. *Science* **234:**1005–1009.

Bollini, R., and M. J. Chrispeels. 1978. Characterization and subcellular localization of vicillin and phyto-hemaglutinin, the two major reserve proteins of *Phaseolus vulgaris. Planta* **142:**291–298.

Chen, Z. L., N. S. Pan, and R. N. Beachy. 1988. A DNA sequence element that confers seed-specific enhancement of a constitutive promoter. *EMBO J.* **7:**297–302.

Chen, Z. L., M. A. Schuler, and R. N. Beachy. 1986. Functional analysis of regulatory elements in a plant embryo-specific gene. *Proc. Natl. Acad. Sci. (USA)* **83:**8560–8564.

Chou, P. Y., and G. D. Fasman. 1978. Prediction of the secondary structure of proteins from amino acid sequence. *Adv. Enzymol.* **47:**45–148.

Colot, V., L. S. Robert, T. A. Kavanagh, M. W. Beavan, and R. D. Tompson. 1987. Localization of sequences in wheat endosperm protein genes which confer tissue-specific expression in tobacco. *EMBO J.* **6:**3559–3564.

Creighton, T. E. 1984. *Proteins.* New York: Freeman.

Crouch, M., K. Tenberge, N. E. Simone, and R. Ferl. 1983. Sequence of the 1.7K storage protein of *Brassica napus. Mol. Appl. Genet.* **2:**273–283.

Degrado, W. F., and J. D. Lear. 1985. Induction of peptide conformation at apolar/water interfaces. *J. Am. Chem. Soc.* **107:**7684–7689.

Degrado, W. F., Z. R. Wasserman, and J. D. Lear. 1989. Protein design, a minimalist approach. *Science* **241:**622–628.

Esen, E. 1986. Separation of alcohol-soluble proteins (zeins) from maize into three fractions by differential solubility. *Plant Physiol.* **80:**623–627.

Fasman, G. 1989. Protein conformational prediction. *Trends Biochem. Sci.* **14:**295–299.

Finley, D., and A. Varshavsky. 1985. The ubiquitin system: Functions and mechanisms. *Trends Biochem. Sci.* **10:**343–346.

Forde, B. G., M. Kreis, M. S. Williamson, R. P. Fry, and J. Pywell. 1985. Short tandem repeats shared by B- and C-hordein cDNAs suggest a common evolutionary origin for two groups of cereal storage protein genes. *EMBO J.* **4:**9–15.

Goldberg, A. L., and A. C. St. John. 1976. Intracellular protein degradation in mammalian and bacterial cells: part 2. *Ann. Rev. Biochem.* **45:**747–803.

Greenwood, J. S., and M. J. Chrispeels. 1985. Correct targeting of the bean storage protein phaseolin in the seeds of transformed tobacco. *Plant Physiol.* **79:**65–71.

Gross, D. S., and W. T. Garrard. 1987. Poising chromatin for transcription. *Trends Biochem.* **12:**293–296.

Ho. S. P., and W. F. Degrado. 1987. Design of a 4-helix bundle protein: Synthesis of peptides which self-associate into helical protein. *J. Am. Chem. Soc.* **109:**6751–6758.

Hoffmann, L. E., D. D. Donaldson, and E. M. Herman. 1988. A modified storage protein is synthesized, processed, and degraded in the seeds of transgenic plants. *Plant Mol. Biol.* **11:**717–729.

Hoffmann, L. E., D. D. Donaldson, R. Bookland, K. Rashka, and E. M. Herman. 1987. Synthesis and protein body deposition of maize 15-kd zein in transgenic tobacco seeds. *EMBO J.* **6:**3213–3221.

Hol, W. G., and H. C. Sanders. 1981. Dipole of the α-helix and β-sheet: Their role in protein folding. *Nature* **294:**532–536.

Jaynes, J. M., P. Nagpala, L. Destefano, T. Denny, C. Clark, and J.-H. Kim. 1992. Expression of a de novo designed peptide in transgenic tobacco plants confers enhanced resistance to *Pseudomonas solanacearum* infection. Submitted to *Proc. Natl. Acad. Sci. (USA)*

Jaynes, J. M., M. S. Yang, N. O. Espinoza, and J. H. Dodds. 1986. Plant protein improvement by genetic engineering: Use of synthetic genes. *Trends Biotechnol.* **4:**314–320.

Jones, J. D. G., and D. E. Gilbert. 1987. T-DNA structure and gene expression in petunia plants transformed by *Agrobacterium tumefaciens* C58 derivatives. *Mol. Gen. Genet.* **207:**478–485.

Kabsch, W., and C. Sander. 1983. How good are predictions of protein structure? *FEBS Lett.* **155:**179–182.

Kane, J. F., and D. L. Hartley. 1988. Formation of recombinant protein inclusion bodies in *Escherichia coli. Trends Biotechnol.* **6:**95–101.

Kasarda, D. D., T. W. Okita, J. E. Bernardin, P. A. Baecker, and C. C. Nimmo. 1984. DNA and amino acid sequences of alpha and gamma gliadins. *Proc. Natl. Acad. Sci. (USA)* **81:**4712–4716.

Keris, M., P. R. Shewry, B. G. Forde, G. Forde, and J. Miflin. 1985. Structure and evolution of seed storage proteins and their genes with particular reference to those of wheat, barley and rye. *Oxford Survey Plant Mol. Cell Biol.* **2:**253–317.

Komoriya, A., and J. M. Chaiken. 1982. Sequence modeling using semisynthetic ribonuclease S. *J. Biol. Chem.* **257:**2599–2604.

Larkins, B. A. 1983. Genetic engineering of seed storage protein. In *Genetic Engineering of Plants,* ed. B. A. Larkins, pp. 93–120. New York: Plenum.

Larkins, B. A., K. Pederson, M. D. Mark, and D. R. Wilson. 1984. The zein protein of maize endosperm. *Trends Biochem. Sci.* **9:**306–308.

Lawrence, M. C., E. Suzuki, J. N. Varghes, P. C. Davis, A. Van Donkelaar, P. A. Tulloch, and P. M. Collman. 1990. The three-dimensional structure of the seed storage protein phaseolin at 3 Å resolution. *EMBO J.* **9:**9–15.

Lear, J. D., Z. R. Wasserman, and W. F. Degrado. 1988. Synthetic amphiphilic peptide model for protein ion channels. *Science* **240:**1177–1181.

Lending, C. R., A. Kriz, B. A. Larkins, and C. E. Bracker. 1988. Structure of maize protein bodies and immunocytochemical localization of zeins. *Protoplasma* **143:**51–62.

Lycett, G. W., R. D. Cory, A. H. Shirsat, D. M. Richards, and D. Boulter. 1985. The 5'-flanking regions of three pea legumin genes: Comparison of DNA sequences. *Nucleic Acids Res.* **13:**6733–6743.

Marqusee, S., and R. Baldwin. 1987. Helix stabilization by GLU-LYS salt bridges in short peptides of de novo design. *Proc. Natl. Acad. Sci. (USA)* **84:**8898–8902.

Marries, C., P. Gallois, J. Copley, and M. Keris. 1988. The 5'-flanking region of a barley B hordein gene controls tissue and developmental specific CAT expression in tobacco plants. *Plant Mol. Biol.* **10:**359–366.

Mutter, M. 1988. Nature's rules and chemist's tools: A way for creating novel proteins. *Trends Biochem. Sci.* **13:**260–264.

Neurath, H. 1989. Proteolytic processing and physiological regulation. *Trends Biochem. Sci.* **14:**268–271.

Okamuro, J. K., K. D. Jofuku, and R. B. Goldberg. 1986. Soybean seed lectin gene and flanking nonseed protein genes are developmentally regulated in transformed tobacco plants. *Proc. Natl. Acad. Sci. (USA)* **83:**8240–8244.

Pace, C. N., and A. J. Barret. 1984. Kinetics of tryptic hydrolysis of the arginine-valine bond in folded and unfolded ribonuclease T1. *Biochem. J.* **219:**411–417.

Pakula, A. A. and R. T. Sauer. 1986. Bacteriophage 1 Cro mutation: Effect on activity and intracellular degradation. *Proc. Natl. Acad. Sci. (USA)* **82:**8829–8833.

Pakula, A. A., and R. T. Sauer. 1989. Amino acid substitutions that increase the thermal stability of the I Cro protein. *Proteins* **5:**202–210.

Parasell, D. A., and R. T. Sauer. 1989. The structural stability of a protein is an important determinant of its proteolytic susceptibility in *Escherichia coli. J. Biol. Chem.* **264**:7590–7595.

Pederson, K., P. Agros, S. V. L. Naravana, and B. A. Larkins. 1986. Sequence analysis and characterization of a maize gene encoding a high-sulfur zein protein of Mw 15,000. *J. Biol. Chem.* **201**:6279–6284.

Pernollet, J. C., and J. Mosse. 1983. Structure and location of legume and cereal seed storage protein. *Seed Proteins (Phytochem. Soc. Europe Symp. Series)* **20**:155–187.

Presnell, S. R., and F. E. Cohen. 1989. Topological distribution of a four-α-helix bundle. *Proc. Natl. Acad. Sci. (USA)* **86**:6592–6596.

Presta, L. G., and G. D. Rose. 1988. Helix signals in proteins. *Science* **240**:1632–1641.

Rafalski, J. A., K. Scheets, M. Metzler, and D. M. Peterson. 1984. Developmentally regulated plant genes: The nucleotide sequence of a wheat gliadin geonomic clone. *EMBO J.* **3**:1409–1415.

Richardson, J. S., and D. C. Richardson. 1988. Amino acid preferences for specific locations at the ends of α-helices. *Science* **240**:1648–1652.

Richardson, J. S., and D. C. Richardson. 1989. The de novo design of protein structures. *Trends Biochem. Sci.* **14**:304–309.

Sanders, P. R., J. A. Winter, A. R. Barnason, and S. G. Rogers. 1987. Comparison of cauliflower mosaic virus 35S and nopaline synthetase promoters in transgenic plants. *Nucleic Acids Res.* **15**:1543–1558.

Scheraga, H. 1978. Use of random copolymers to determine helix-coil stability constants of the naturally occurring amino acids. *Pure Appl. Chem.* **50**:315–324.

Scheraga, H. A. 1985. Effect of side chain-backbone electrostatic interaction on the stability of α-helices. *Proc. Natl. Acad. Sci. (USA)* **82**:5585–5587.

Scott, R. J., and J. Draper. 1987. Transformation of carrot tissue derived from proembryogenic suspension cells: A useful model system for gene expression studies in plants. *Plant Mol. Biol.* **8**:265–274.

Sengupta, G. C., N. A. Reichert, R. F. Baker, T. C. Hall, and J. D. Kemp. 1985. Developmentally regulated expression of the bean β-phaseolin gene in tobacco seed. *Proc. Natl. Acad. Sci. (USA)* **82**:3320–3324.

Shen, S.-H. 1984. Multiple joined genes prevent product degradation in *E. coli. Proc. Natl. Acad. Sci. (USA)* **81**:4627–4631.

Shoemaker, K. R., P. S. Kim, E. J. York, J. M. Stewart, and R. L. Baldwin. 1987. Test of helix dipole model for stabilization of α-helices. *Nature* **326**:563–566.

Staswick, P. E. 1989. Preferential loss of an abundant storage protein from soybean pods during seed development. *Plant Physiol.* **90**:1251–1255.

Stockhaus, J., P. Eckes, A. Blau, J. Schell, and L. Willmitzer. 1987. Organ-specific and dosage-dependent expression of a leaf/stem specific gene from potato after tagging and transfer into potato and tobacco plants. *Nucleic Acids Res.* **15**:3479–3491.

Sueki, M., S. Lee, S. P. Power, J. B. Denton, Y. Konishi, and H. Scheraga. 1984. Helix-coil stability constants for the naturally occurring amino acids in water. *Macromolecules* **17**:148–155.

Twell, D., and G. Ooms. 1987. The 5'-flanking DNA of a patatin gene directs tuber specific expression of a chimeric gene in potato. *Plant Mol. Biol.* **9**:365–375.

Wallace, J. C., G. Galili, E. E. Kawata, R. E. Cuellar, M. A. Shotwell, and B. A. Larkins. 1988. Aggregation of lysine containing zeins into protein bodies in *Xenopus* oocytes. *Science* **240**:662–664.

Weiler, E. W., and J. Schroder. 1987. Hormone genes and crown gall disease. *Trends Biochem. Sci.* **12**:271–275.

Wenzler, H. C., G. A. Mignery, L. M. Fisher, and W. D. Park. 1989. Analysis of a chimeric class I potatin-GUS gene in transgenic potato plants: High level expression of tubers and sucrose-inducible expression in cultured leaf and stem explants. *Plant Mol. Biol.* **12**:41–50.

Yang, M. S., N. O. Espinoza, J. H. Dodds, and J. M. Jaynes. 1989. Expression of a synthetic gene for improved protein quality in transformed potato plants. *Plant Sci.* **64**:99–111.

Zimm, B. H., and J. R. Bragg. 1959. Theory of the phase transition between helix and random coil in polypeptide chains. *J. Chem. Phys.* **31**:526–535.

2

Deamidation and Phosphorylation to Improve Protein Functionality in Foods

Frederick F. Shih, Jamel S. Hamada, and Wayne E. Marshall

Many proteins utilized for human consumption require structural modification to achieve the proper functional properties for use as food ingredients. Food protein modification is normally accomplished by either chemical or enzymatic methods. Chemical hydrolysis with acid or base and enzymatic proteolysis have been popular and useful modification techniques used by the food processing industry. However, existing commercial modification procedures are limited in number and usefulness. Other chemical and enzymatic modification methods must be developed and made available to the food processor, particularly methods that do not significantly decrease the nutritional value of the protein.

Excellent reviews have appeared that give many examples of chemical (Feeney 1977; Meyer and Williams 1977; Feeney, Yamasaki, and Geoghegan 1982; Feeney and Whitaker 1985) and enzymatic (Whitaker 1977; Whitaker and Puigserver 1982; Feeney and Whitaker 1985) methods for food protein modification. Since the end use of modified food protein is incorporation into food products, there are limitations placed on the modified product if it has not been proved safe for human consumption. In addition to restrictions imposed by the Food and Drug Administration (FDA), there are further restrictions imposed by the food industry itself. Since most food proteins are consumed in large quantities, the modified proteins must be produced in bulk with anticipated small profit margins. Therefore, industrial modification processes must be specific, reproducible, and cost effective. Given these demanding requirements, most protein modification methods remain laboratory curiosities.

A persistent roadblock to the expanded use of food proteins, especially soy protein, is poor solubility and emulsification under mildly acidic (pH 3-6) conditions (Hirotsuka et al. 1984), which excludes its use in coffee whitener, acidic beverages, pourable, and nonpourable dressings. Our review will focus on modifications that can improve the solubility and emulsification of food proteins by imparting additional negative charges to change the isoelectric pH of the protein to a lower value. In our opinion, the most promising routes to commercially increasing protein negative charge are through deamidation and phosphorylation. These modification methods have received considerable attention in our as well as other laboratories over the past few years, in part because they do not reduce protein nutritional value. The following discussion will be limited to these two topics. The emphasis will be on enzymatic methods developed to modify soy protein, but not to the total exclusion of chemical modification. However, enzymatic processes are more likely to pass regulatory hurdles than chemical treatments. Enzymes are generally more site specific than chemical reagents and are much less likely to produce unwanted side reactions. Through protein engineering, large quantities of needed enzymes with appropriate physical, chemical, and catalytic properties can be produced at reasonable cost, thus keeping the modification process cost effective.

DEAMIDATION OF PROTEINS

Protein deamidation refers to the conversion of peptide-bound, pendant amide groups to carboxylic acid groups with the concomitant release of ammonia. While very few mechanisms for enzymatic deamidation have been reported in the literature, pathways of nonenzymatic deamidation have been extensively investigated. The conversion products for chemical deamidation appear to vary widely depending on factors including pH, temperature, buffer composition, and sequence and size of amino acids in the deamidated substrate (Bhatt, Patel, and Borchardt 1990; Patel and Borchardt 1991a; Patel and Borchardt 1990b). Particularly, pH appears to exert major control over the nonenzymatic deamidation of asparaginyl residues in polypeptides. At alkaline and neutral pH's, deamidation of asparagine (Asn) residues was reported to involve the formation of an intramolecular cyclic imide intermediate (Fig. 2-1, Pathway I) (Aswad 1984; Meinwald, Stinson, and Scheraga 1986; Geiger and Clarke 1987). The intermediate could be hydrolyzed to a normal α-linked aspartate (asp) residue and a β-linked Asp residue (iso-Asp) with the latter normally predominating (Sondheimer and Holley 1954; Battersby and Robinson 1955; Lura and Schirch 1988). Under acidic conditions, the reaction was reported to proceed via direct hydrolysis of Asn residues to Asp residues without the formation of

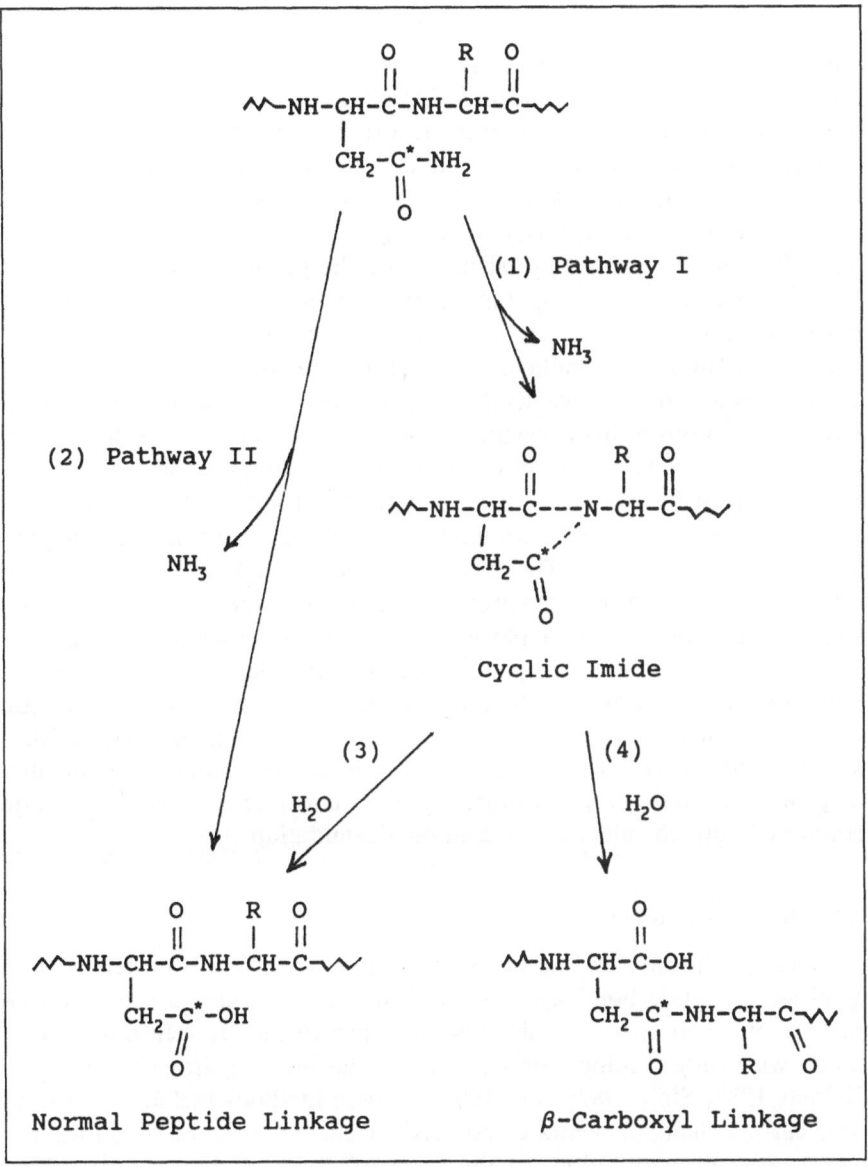

FIGURE 2-1.   Mechanism of deamidation: (1) Pathway I, formation of a cyclic imide intermediate and release of ammonia; (2) Pathway II, direct hydrolysis of asparagine (Asn), release of ammonia, and formation of aspartic (Asp); (3) hydrolysis of the cyclic imide intermediate to form an α-carboxyl linkage; (4) hydrolysis of the cyclic imide intermediate to form a β-carboxyl linkage. The asterisk marks the position of the side-chain carbonyl carbon. In the formation of a β-carboxyl linkage (iso-Asp), this carbon becomes part of the main chain and increases the chain length by one carbon.

a cyclic imide intermediate (Fig. 2-1, Pathway II) (Bhatt, Patel, and Borchardt 1990). This straightforward deamidation under acidic conditions is significant because chemical deamidation of food proteins, normally conducted in mildly acidic solutions, is expected to produce the normal α-linked carboxylic residues instead of the nutritionally undesirable β-linked isomers. The ease of deamidation depends both on the amino acid sequence flanking the carboxamide group (Geiger and Clarke 1987; Harding 1985) and on the tertiary structure of the protein (Kossiakoff 1988). While both asparagine and glutamine can undergo deamidation, asparagine has been shown to deamidate at much higher rates (Robinson, Scotchler, and McKerrow 1973; Sondheimer and Holley 1954).

Proteins are most conveniently deamidated by acid or base hydrolysis. However, protons or hydroxyl ions catalyze the hydrolysis of both carboxamide and peptide bonds, resulting in deamidated and hydrolyzed proteins. Extensive protein hydrolysis is undesirable for food use because it could result in reduced protein functionality and the release of bitter-tasting peptides (Adler-Nissen 1986). The problem can be avoided by developing methods that selectively deamidate proteins and minimize peptide bond hydrolysis. Deamidation can be a particularly useful modification method for vegetable proteins such as soy and wheat proteins because they are rich in glutamine and asparagine. Therefore, even low-level (2–5%) deamidation could result in significant improvement of their functional properties (Matsudomi et al. 1985). Food scientists are beginning to recognize the potential of protein deamidation, and more reports are available in recent years on studies of both chemical and enzymatic deamidation.

Chemical Deamidation

As noted previously, acid- or base-catalyzed deamidation of proteins causes significant peptide bond hydrolysis. However, methods have been developed in our laboratory in which vegetable proteins are significantly deamidated with only a minor amount of peptide bond hydrolysis (Shih and Kalmar 1987; Shih 1987; Shih 1990c). These methods include the use of long carbon chain alkylsulfate and alkylsulfonate anions (Shih and Kalmar 1987), arylsulfonate anions in the form of cation exchange resins (Shih 1987), and common anions such as phosphate or bicarbonate as catalysts (Shih 1991). The deamidated proteins, with up to 40% protein deamidation and only 1–4% peptide bond hydrolysis, showed substantially improved solubility, water-binding capacity, foam expansion, emulsion capacity, and emulsion viscosity over their unmodified counterparts (Shih and Kalmar 1987; Shih 1987).

On the basis of his earlier work, Shih developed a patented process (Shih

1989) for protein deamidation that uses long chain alkylsulfate, alkanesulfonate, and arylsulfonate catalysts under conditions that minimize peptide bond hydrolysis. The process is amenable to industrial scaleup and should be cost effective. In fact, soy proteins, chemically deamidated by this process, have been evaluated in liquid and dry coffee whitener formulations with considerable improvement in whitening power and a reduction in feathering when compared to unmodified soy (Shih, unpublished results).

Enzymatic Deamidation

Enzymatic deamidation could be a viable alternative to chemical deamidation because enzymes are substrate specific, that is, no peptide bond hydrolysis, and reactions are catalyzed at mild temperatures (30–40°C) and neutral pH. The most significant deamidating enzymes, based on the recent literature, are deamidating proteases, transglutaminase, and peptidoglutaminase. The following discussions involve protein deamidation using these enzymes.

Deamidation by Proteases

Kato et al. (1987*a*; 1987*b*) reported using proteases for the deamidation of food proteins. About 20% of the amide bonds were hydrolyzed using the proteases papain, pronase E, or chymotrypsin at pH 10 and 20°C. Peptide bond hydrolysis ranged from 0% to 8%, with soy protein and gluten receiving the greatest degree of hydrolysis. Kato, Lee, and Kobayashi (1989) later reported deamidation of selected plant and animal proteins with chymotrypsin immobilized on controlled pore glass. The reaction was carried out at pH 10 and 20°C and resulted in a range of deamidation values from 5% to 10%. Kato et al. (1987*a*; 1987*b*) and Kato, Lee, and Kobayashi (1989) investigated the functional properties of proteins deamidated by proteases and found an increase in gluten solubility at all pH values in the range of 2 to 12. Emulsifying and foaming properties of the proteins were increased by treatment with immobilized chymotrypsin.

Enzymatic deamidation using commercial food-grade proteases appears to be a straightforward and cost effective process for producing modified food proteins. However, recent studies by Shih (1990*a*; 1990*b*) raise serious questions about the use of proteases to deamidate food proteins. Shih used reaction conditions very similar to those of Kato et al. (1987*a*; 1987*b*) and observed that none of the proteases investigated, including papain, chymotrypsin, and pronase E showed any deamidating activity toward soy protein. Instead, Shih found substantial proteolysis, and the generation of ammonia. The ammonia was from nonenzymatic deamidation of free glutamine and deamination of certain free amino acids in the hydrolyzates.

The nonenzymatic deamidation of free glutamine was most likely catalyzed by phosphate and bicarbonate anions in the buffer (Shih 1990a). Thus, the improvement of protein functional properties as described by Kato et al. (1987a; 1987b) and Kato, Lee, and Kobayashi (1989) may or may not be the result of enzymatic deamidation. Further research is needed to explore the exact role of proteases in protein deamidation.

Deamidation by Transglutaminase

Lorand and Conrad (1984) have reviewed various aspects of transglutaminases (TGases, EC 2.3.2.13) emphasizing the biological role of these enzymes. A Ca^{++}-activated enzyme derived from the soluble fraction of guinea pig liver, TGase normally catalyzes the incorporation of a number of primary amines into proteins and polypeptides through an acyl transfer reaction between the γ-carbonyl group in glutamine and a receptor amino group (Fig. 2-2). If the primary amine is part of the same or another protein, intra- or intermolecular cross-linking occurs. If no primary amine is present, or if a primary amine is present but blocked, water can act as an acyl acceptor, and deamidation occurs with the release of ammonia to give peptide-bound glutamic acid (Fig. 2-2).

Prior to TGase catalysis, proteins have been acylated, deaminated, or guanidinated to avoid cross-linking (Mycek and Waelsch, 1960; Bercovici, Gaertner, and Puigserver 1987). If proteins are acylated with citraconic anhydride, then the blockage is reversible and the citraconylated residue can be regenerated under acidic conditions (Brinegar and Kinsella 1980). As an example, Motoki et al. (1986) used guinea pig TGase to catalyze the deamidation of α_{-s1}-casein according to the scheme shown in Figure 2-3. Citraconylation was performed prior to deamidation. Almost complete citraconylation was observed and about 80% of the glutamine was deamidated. After removal of the blocking groups, the deamidated protein was found to have improved solubility, particularly in the pH range between 3 and 5.

Since food protein must be chemically modified prior to TGase deamidation, the chemical pretreatment poses potential problems for the food industry. If irreversible blocking agents such as succinic or acetic anhydride are used, then these reagents must meet FDA approval as food additives. However, since lysine is the preferred target for these blocking agents, the utilization of lysine would be reduced (Bjarnason and Carpenter 1969) and the nutritional quality of the protein would diminish. If the reversible blocking reagent citraconic anhydride is used, then its hydrolyzed product, citraconic acid, would have to be separated from the modified protein before being used in foods. Also, the blocking step would probably add prohibitive cost to the modification process. The use of TGase in a scaleup, industrial deamidation process appears unfeasible at present.

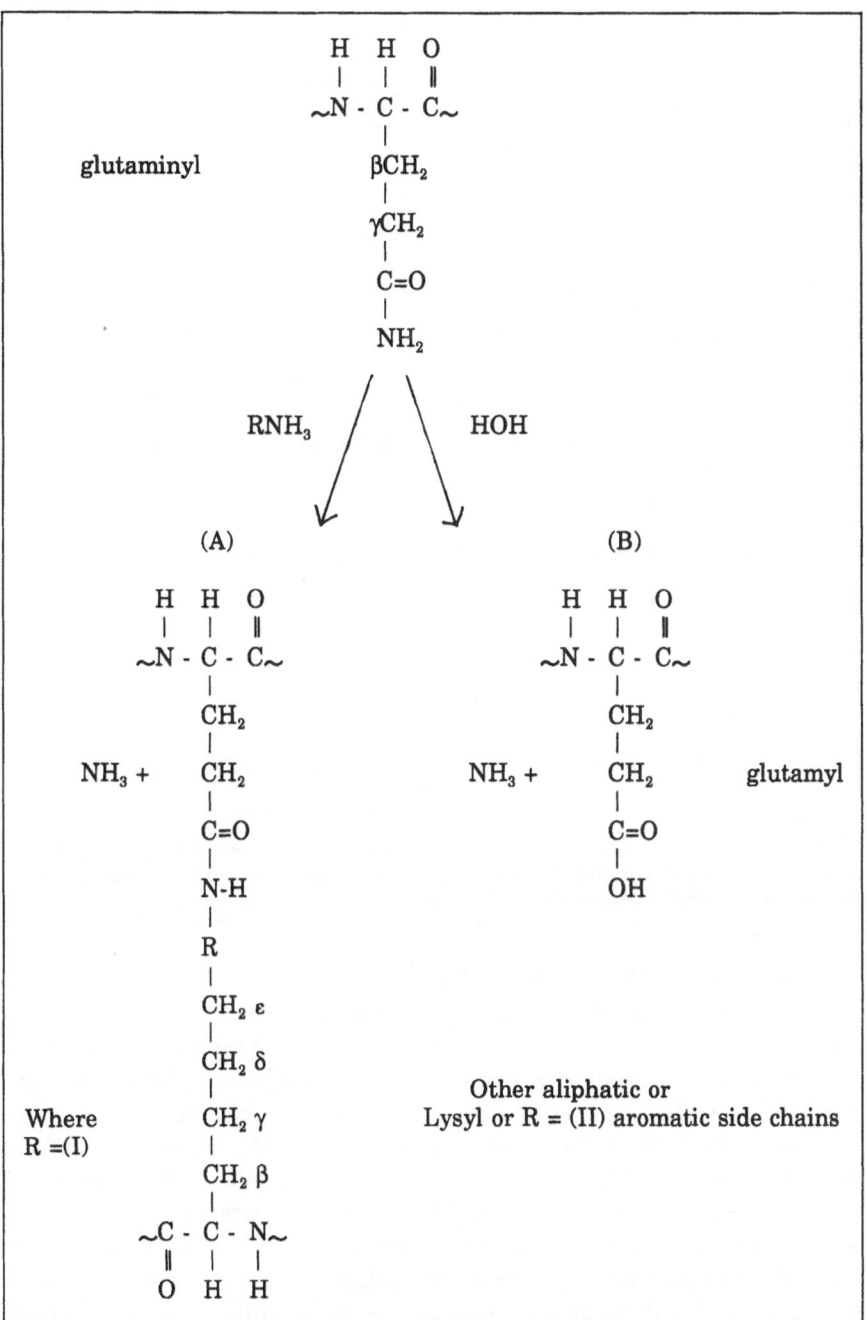

FIGURE 2–2. General reactions catalyzed by transglutaminase. Structure (*A*) is formed by an acyl-transfer reaction in which the γ-carboxamide group of peptide-bound glutamine acts as an acyl donor and reacts with primary amino groups of a variety of compounds, including the ε-amino group of a peptide-bound lysine residue, which act as an acyl acceptor. The reaction forms monosubstituted γ-amides of peptide-bound glutamic acid (structures A.I and A.II). Structure (*B*) is formed when water is the acyl acceptor and deamidation occurs in the absence of amines to give peptide-bound glutamic acid.   (From Folk and Finlayson 1977)

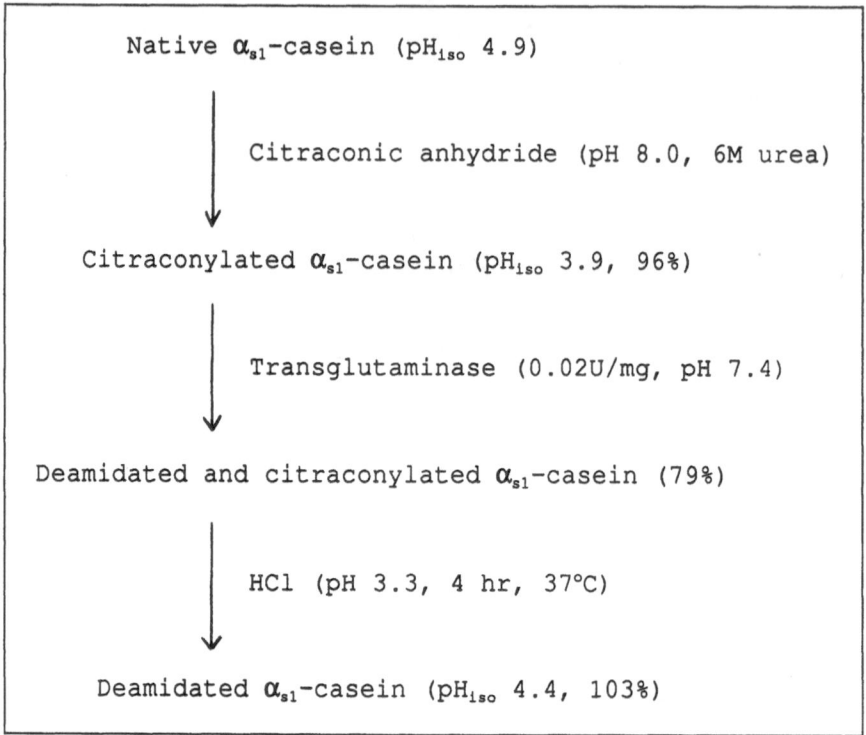

Native α_{s1}-casein (pH$_{iso}$ 4.9)

Citraconic anhydride (pH 8.0, 6M urea)

Citraconylated α_{s1}-casein (pH$_{iso}$ 3.9, 96%)

Transglutaminase (0.02U/mg, pH 7.4)

Deamidated and citraconylated α_{s1}-casein (79%)

HCl (pH 3.3, 4 hr, 37°C)

Deamidated α_{s1}-casein (pH$_{iso}$ 4.4, 103%)

FIGURE 2–3.   A flow diagram showing the deamidation of α_{s1}-casein by transglutaminase. pH$_{iso}$ refers to the protein isoelectric pH and the percentages indicate the degree of modification.   (From Motoki et al. 1986)

Deamidation by Peptidoglutaminase

Enzymes capable of deamidating glutamine to glutamate are L-glutamine amido hydrolases (EC 3.5.1.2). Kikuchi et al. (1971) reported L-glutamine amido hydrolases capable of deamidating peptide-bound glutamine. They obtained an L-glutamine amido hydrolase from the soil bacterium *Bacillus circulans,* and gave it the generic name peptidoglutaminase (PGase). PGase was found to deamidate the carboxamide of glutamine, where the glutamine was part of a small peptide. Their PGase preparation contained two distinct enzyme activities, whcih they labeled PGase I and II. PGase I catalyzed the deamidation of C-terminal glutamine in small peptides, while PGase II catalyzed the deamidation of glutamine within the peptide chain. Gill et al. (1985) detected limited deamidating activity of PGase II toward casein and whey protein hydrolyzates.

Hamada et al. (1988) also investigated the potential of a partially purified PGase preparation (mixture of PGases I and II) for soy protein modification. They observed that PGase readily deamidated glutamine in soy protein hydrolyzates, but its activity toward intact soy protein was small. Hamada (1990a) developed a methodology for PGase deamidation of a variety of animal and plant proteins and their hydrolyzates (Table 2-1). Enzyme activity toward intact protein substrates varied from no modification in egg albumin to 6% deamidation in soy protein. However, deamidation of hydrolyzates was extensive. The rsults in Table 2-1 show that PGase readily deamidates glutamine in protein hydrolyzates and that the use of proteolysis prior to PGase treatment substantially enhanced deamidation.

Hamada and Marshall (1988) found that, in addition to proteolysis, heating soy protein solutions also enhanced PGase deamidation. In fact, a combination of heat and proteolysis was particularly effective. Hamada and Marshall (1988) quantified the effects of heat and proteolysis for PGase activity toward soy protein, as shown in Figure 2-4. The most effective treatment consisted of heat applied before and after hydrolysis (Fig. 2-4D), and at 20% DH (degree of hydrolysis), 28% of the amide groups in the hydrolyzed protein were deamidated. The functional properties of soy proteins subjected to heat and hydrolysis were also examined (Hamada and Marshall 1989). The solubility of deamidated protein increased at pH 4-7 as well as under alkaline conditions. Deamidation also increased emulsifying

TABLE 2-1   PGase Deamidation of Food Proteins and Protein Hydrolysates

Protein or Protein Hydrolysate	N.F.	% Protein	Free NH_3 mmoles/g Protein	Total Amide mmoles/g Protein	% PGase Deamidation[b]
Water extract of soy flakes	6.25	56.9	0.026	1.00	6.4
Soy peptone (type IV)[c]	6.25	58.5	0.032	0.68	56.0
Papainic soy hydrolysate[d]	6.25	80.6	0.036	0.90	14.5
Wheat gluten[c]	5.70	73.0	0.018	2.50	1.3
Corn gluten[c]	5.60	56.6	0.032	1.23	2.8
Corn gluten hydrolysate[c]	5.60	54.9	0.174	1.22	30.9
Rice protein hydrolysate[d]	5.95	82.3	0.163	1.38	13.7
Egg albumin[c]	6.25	82.5	0.017	0.86	0.0
Papinic egg albumin hydrolysate[d]	6.25	76.3	0.036	0.65	39.6
Lactalbumin hydrolysate (DC-A)[d]	6.38	78.5	0.121	0.71	39.0
Lactalbumin hydrolysate (DC-K)[d]	6.38	61.9	0.146	0.85	37.4
Pancreatinic casein hydrolysate[d]	6.38	79.8	0.099	0.88	30.4
Casein hydrolysate (DC-L)[d]	6.38	70.8	0.249	1.12	27.2
Casein hydrolysate (DC-N)[d]	6.38	79.1	0.450	0.95	26.3

Source: From Hamada (1990a).

[a] % Protein = %N × N.F.
[b] Calculated from values of NH_3 released enzymatically and total amide contents.
[c] From Sigma Chemical Co. (St. Louis, Mo.).
[d] Enzymatic protein hydrolysates from Deltown Chemurgic Co. (Fraser, N.Y.).

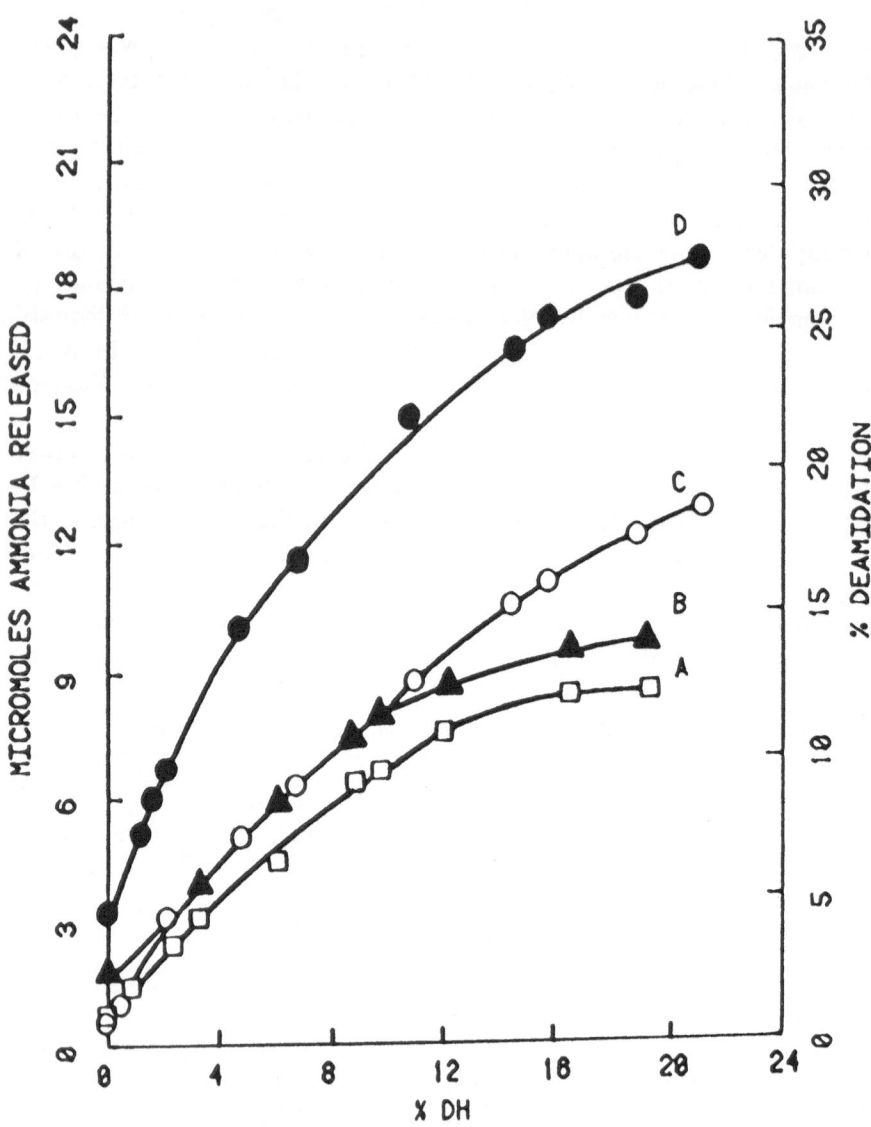

FIGURE 2-4.   Effect of proteolysis and heat treatment on the PGase deamidation of soy protein: (*A*) unheated; (*B*) heated after proteolysis; (*C*) heated before proteolysis, and (*D*) heated before and after proteolysis.   (From Hamada and Marshall 1988)

activity, emulsion stability, and foaming power, but had no apparent effect on foam stability (Hamada and Marshall 1989). The improved functionality of deamidated soy proteins has lead us to consider using PGase in an enzyme reactor for the continuous deamidation of food proteins.

Hamada (1989) investigated the suitability of ultrafiltration membranes with specific pore sizes for use as PGase containment vessels. Later, Hamada (1990b) developed a batch membrane bioreactor method where hydrolyzed protein substrate can be circulated through the containment vessel to achieve complete deamidation. This method can be used in designing an efficient, continuous deamidation process using membrane technology. Hamada is currently developing a PGase purification procedure using preparative high-performance liquid chromatography (HPLC) to obtain large quantities of high specific activity PGase at protentially reasonable cost (unpublished observations). This recent research greatly improves the prospects of using PGase for protein and peptide deamidation on an industrial scale.

PHOSPHORYLATION OF PROTEINS

Inorganic phosphate (P_i) can be transferred to proteins by either O- or N-esterification reactions. In an O-esterification reaction, P_i reacts with the primary or secondary hydroxyls on serine or threonine, respectively; or with the weakly acidic hydroxyl on tyrosine, forming a $-C-O-P_i$ bond (Fig. 2-5). In N-esterification, P_i combines with the ϵ-amino group of lysine, the imidazole group of histidine, or the guanidino group of arginine, forming a $-C-N-P_i$ bond (Fig. 2-5). The nitrogen-bound phosphates are acid labile and are readily hydrolyzed at pH values at or below 7 (Matheis et al. 1983). Proteins containing oxygen-bound phosphate are acid stable and are the modification of choice for food proteins (Matheis and Whitaker 1984), since the pH of most foods is 3-7.

Chemical Phosphorylation

Matheis and Whitaker (1984) have reviewed methods to chemically phosphorylate food proteins. Of the phosphorylating agents surveyed, only phosphorus oxychloride ($POCl_3$) and sodium trimetaphosphate (STMP), in their opinion, might prove economical and practical for large-scale phosphorylation of food proteins. Hirotsuka et al. (1984) phosphorylated lysine and histidine residues on soy protein isolate with $POCl_3$. The modified soy protein exhibited increased solubility and emulsifying activity, and was stable above pH 3.5. However, rapid dephosphorylation was observed below 3.5. Woo, Creamer, and Richardson (1982), Woo and Richardson (1983),

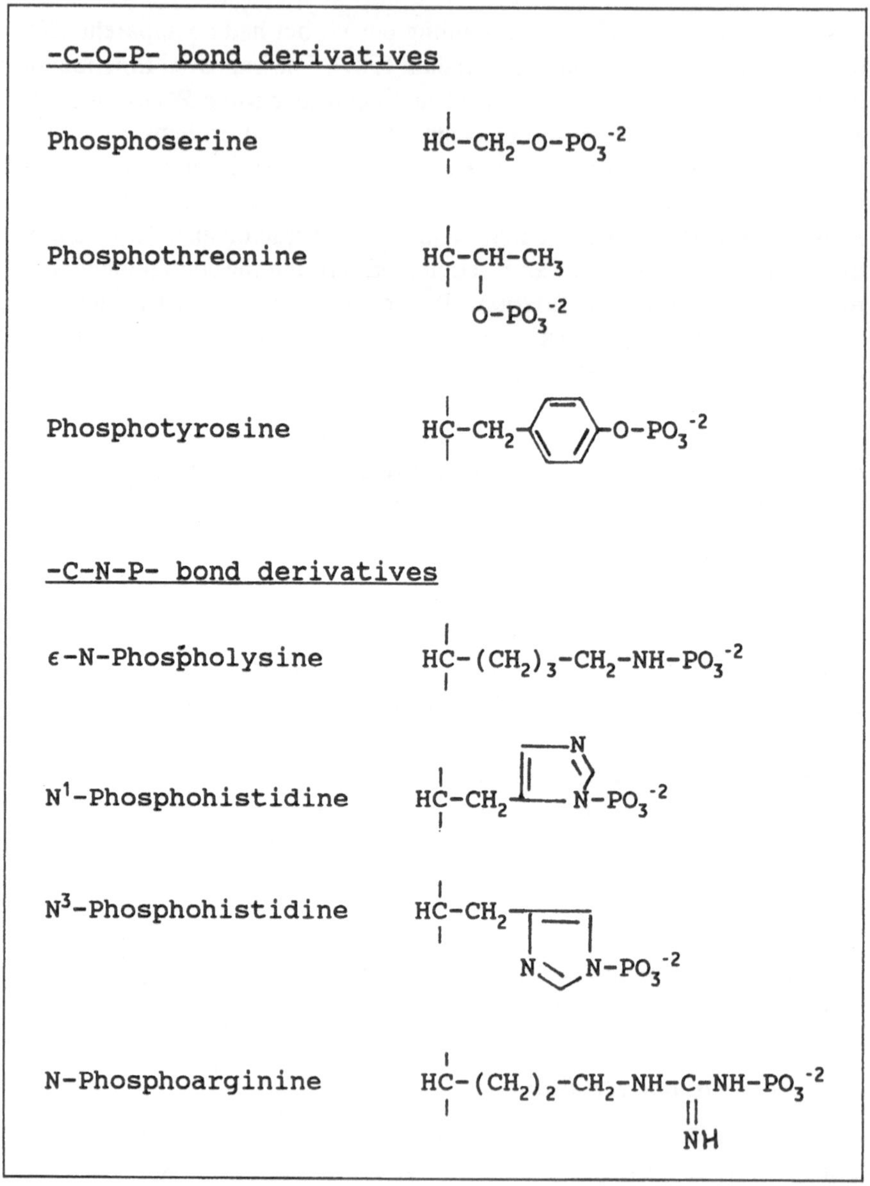

-C-O-P- bond derivatives

Phosphoserine

$$\text{H}\overset{|}{\underset{|}{\text{C}}}\text{-CH}_2\text{-O-PO}_3^{-2}$$

Phosphothreonine

$$\text{H}\overset{|}{\underset{|}{\text{C}}}\text{-CH-CH}_3$$
$$\qquad\qquad\underset{\text{O-PO}_3^{-2}}{|}$$

Phosphotyrosine

$$\text{H}\overset{|}{\underset{|}{\text{C}}}\text{-CH}_2\text{—}\langle\!\!\!=\!\!\!\rangle\text{-O-PO}_3^{-2}$$

-C-N-P- bond derivatives

ϵ-N-Phospholysine

$$\text{H}\overset{|}{\underset{|}{\text{C}}}\text{-(CH}_2)_3\text{-CH}_2\text{-NH-PO}_3^{-2}$$

N^1-Phosphohistidine

$$\text{H}\overset{|}{\underset{|}{\text{C}}}\text{-CH}_2\text{-N-PO}_3^{-2}$$

N^3-Phosphohistidine

$$\text{H}\overset{|}{\underset{|}{\text{C}}}\text{-CH}_2\text{-N-PO}_3^{-2}$$

N-Phosphoarginine

$$\text{H}\overset{|}{\underset{|}{\text{C}}}\text{-(CH}_2)_2\text{-CH}_2\text{-NH-}\overset{}{\underset{\|}{\text{C}}}\text{-NH-PO}_3^{-2}$$
$$\qquad\qquad\qquad\qquad\qquad\text{NH}$$

FIGURE 2-5.   Phosphorylated amino acid derivatives in proteins.

and Matheis et al. (1983) also used $POCl_3$ to phosphorylate β-lactoglobulin, casein, and lysozyme. Matheis et al. (1983) found phosphorylated casein to have decreased water solubility and emulsifying activity due to protein cross-linking, which also occurred with phosphorylated β-lactoglobulin (Woo, Creamer, and Richardson 1982). In contrast, Woo and Richardson (1983) observed increased emulsifying activity and also increased gel-forming properties in the presence of Ca^{++} with phosphorylated β-lactoglobulin. Huang and Kinsella (1987) used $POCl_3$ to phosphorylate yeast protein, and found the treatment to be effective in the removal of nucleic acid and for enhancing functional properties. The phosphorylated yeast protein showed improvement in emulsifying activity and emulsion stability, and produced stable but weak foams at neutral pH. Huang and Kinsella (1987) noted that the in vitro digestibility of yeast protein was not affected by phosphorylation as previously reported for casein (Matheis et al. 1983) and soy protein (Hirotsuda et al. 1984).

In $POCl_3$ phosphorylated food proteins, Hirotsuka et al. (1984) reported up to 140 moles of P_i incorporated per mole of soy isolate, while Woo, Creamer, and Richardson (1982) reported 14 moles of P_i per mole of β-lactogulobulin, and Matheis et al. (1983) found 7.4 and 6.2 moles P_i per mole of casein and lysozyme, respectively. However, $POCl_3$ reacts rapidly with water to produce P_i and HCl vapor, a noxious gas. To minimize the reaction with water $POCl_3$ is dissolved in an organic solvent, such as carbon tetrachloride, which could cause handling problems in a food protein processing facility. In addition, $POCl_3$ promotes protein cross-linking and its site of phosphorylation is nonspecific, as both O- and N-esterification occur. Given these limitations, we do not find $POCl_3$ to be a practical reagent for large-scale food protein phosphorylation.

Sung et al. (1983) used sodium trimetaphosphate (STMP) to modify serine and lysine in soy isolate under alkaline conditions. About 40% of the total serine residues were reported phosphorylated with no protein cross-linking, and the isolate displayed increased solubility and emulsifying properties, particularly under acidic conditions. However, Matheis et al. (1983) were unable to detect any covalently bound P_i in soy protein and lysozyme using the methods of Sung et al. (1983).

Chemical phosphorylation with STMP can avoid many of the problems associated with $POCl_3$. For example, STMP is an FDA-approved food additive (Ellinger 1972), and hydrolysis in water produces only harmless P_i. STMP does not cause protein cross-linking. However, phosphorylation with STMP occurs at alkaline pH (Sung et al. 1983), which could lead to undesirable alkaline hydrolysis and lysinoalanine formation. STMP shows potential for food protein modification on a commercial scale, but further research is needed to determine the exact incorporation of covalently bound

P_i to avoid alkaline pH conditions and to move its specificity toward O-esterification rather than a mixture of O- and N-esterification.

Enzymatic Phosphorylation

Enzymes that phosphorylate proteins are called *protein kinases* (EC 2.7.1.37), which is a generic term for enzymes that transfer the γ-phosphate of ATP to the hydroxyl groups of serine, threonine, or tyrosine residues in proteins. Enzymatic phosphorylation has been extensively investigated because protein phosphorylation–dephosphorylation is an important mechanism in the regulation of a variety of enzymes and proteins in mammalian cells (Krebs and Beavo 1979; Cohen 1982; Nestler, Walaas, and Greengard 1984; Hunter and Cooper 1985). However, very few reports are available that describe enzymatic phosphorylation for food application. Numerous phosphorylating enzymes have been identified and characterized, but none are currently available in quantities large enough for practical food use.

Based on protein substrate specificity, protein kinases can be generally classified into two major types: histone-type and casein (or phosvitin) -type kinase. Histone-type kinases can be cAMP (or cGMP) -dependent or independent, and are active toward histones, but generally will not phosphorylate acidic proteins. Casein-type kinases are generally cyclic nucleotide independent and will phosphorylate acidic substrates such as casein, phosvitin, and nuclear nonhistone proteins, but not histones. These classifications are based on enzymes characterized from animal tissue (Rubin and Rosen 1975; Langan 1973; Krebs 1986; Hanks, Quinn, and Hunter 1988). Since our interest is vegetable (soy) protein phosphorylation, protein kinases from plant sources, expecially germinating soybeans, may have greater specificity and higher catalytic activity toward soy or other vegetable proteins. Unfortunately, only a few protein kinases have been reported to be present in higher plants. Of interest are those that have been isolated and characterized from wheat germ (Polya and Davies 1982), soybeans (Murray, Guilfoyle, and Key 1978; Putnam-Evans, Harmon, and Cormier 1990), and soybean cotyledons (Gowda and Phillay 1982). A few cAMP (cGMP) -independent protein kinases from soybean seedlings have been reported to specifically phosphorylate lysine-rich histones (Lin and Key 1980; Feller 1989). Our laboratory is currently investigating the ability of protein kinases from germinating soybean seedlings to phosphorylate soy proteins.

Protein Kinase from Bovine Cardiac Muscle

For enzymatic phosphorylation to be a viable modification method for the food processing industry, the enzymes have to be readily available. Probably the protein kinase that comes closest to meeting the availability requirement is the adenosine cyclic $3',5'$-monophosphate-dependent protein kinase (cAMPdPK) from bovine cardiac muscle (Smith et al. 1981; Okuno and Fujisawa 1990). It is one of the few protein kinases that is available from at least one commercial source, namely, Sigma Chemical Company, St. Louis, Mo. cAMPdPK exists as an inactive tetrameric holoenzyme designated R_2C_2, with two identical regulatory (R) and two identical catalytic (C) subunits (Krebs and Beavo 1979). Each R subunit binds two molecules of cAMP, which leads to the dissociation of the holoenzyme and activation of the C subunit (Smith et al. 1981). The amino acid sequence of the enzyme has been determined (Shoji et al. 1981; 1983). The DNA segment that encodes this kinase has been isolated, cloned, and expressed (Uhler et al. 1986; Uhler, Chrivia, and McKnight 1986; Uhler and McKnight 1987). Thus, large quantities of active enzyme could be made economically available if it proves useful in food protein modification.

Food Protein Phosphorylation by cAMPdPK

Ross and Bhatnagar (1989) and Seguro and Motoki (1989) used the C subunit of cAMPdPK from bovine cardiac muscle to phosphorylate soybean proteins. This kinase is most active toward serines that are located one or two amino acid residues away from basic amino acids such as arginine and lysine. The highest specificity is toward the amino acid sequences Arg-Arg-X-Ser and Lys-Arg-X-X-Ser (Krebs and Beavo 1979). The polypeptides of glycinin and the subunits of β-conglycinin contain these amino acid sequences. Seguro and Motoki (1989) reported at least 18 phosphorylation sites on the acidic polypeptides of glycinin. In addition, we have located at least one phosphorylation site on β-conglycinin, but none on the basic polypeptides of glycinin (unpublished observations). Other potential serine phosphorylation sites on these proteins may be uncovered as more sequence data become available.

Optimum reaction conditions were established for phosphorylation of soy protein using cAMPdPK from a commercial source (Ross 1989). These results were compared to soy protein phosphorylation catalyzed by a purified laboratory preparation (Table 2-2) (Ross and Bhatnagar 1989). Two isolates and two purified soy storage proteins were used as substrates. For most of the samples, the heat denatured proteins were phosphorylated to a greater degree than native or non-heat-treated proteins. Glycinin was phosphorylated to a greater extent than β-conglycinin in all cases. This observation is understandable considering the difference in the number of potential phosphorylation sites between the two proteins alluded to earlier. The greatest level of phosphorylation (5.6 moles P_i/mole protein) was observed with denatured glycinin (Table 2-2), and compares favorably with the amount of P_i incorporated into β-lactoglobulin, casein, and lysozyme using $POCl_3$

TABLE 2-2   Moles of Phosphorus Incorporated[a] per Mole of Soy Protein[b] Using Different cAMPdPK Preparations

	β-conglycinin		Glycinin	
Ultrafiltrate isolate				
native	0.16[c]	(1.15)[d]	0.70[c]	(1.27)[d]
denatured	0.83	(0.66)	1.01	(1.38)
Purina Protein 620				
native	0.45	(1.16)	1.14	(1.36)
denatured	0.88	(0.83)	0.53	(2.09)
β-conglycinin				
native	0.28	(0.04)		
denatured	1.20	(0.21)		
Glycinin				
native			0.63	(2.08)
denatured			3.96	(5.57)

Source: From Ross and Bhatnagar 1989; Ross 1989.

[a]The amount of P_i incorporated into β-conglycinin and glycinin was determined by separating the proteins on LDS-PAGE, then counting $^{32}P_i$ labels in the protein bands on the gels using a radiochromatograph.
[b]For calculations, molecular weights of 360,000 and 175,000 were used for glycinin and β-conglycin, respectively.
[c]Values denote P_i incorporation using laboratory-prepared cAMPdPK.
[d]Values in parentheses denote P_i incorporation using commercial cAMPdPK under optimum reaction conditions.

as noted earlier. In a commercial process, soy isolate rather than purified storage proteins would be the substrate. In that case, total P_i incorporation into storage protein (glycinin + β-conglycinin) would vary from 2 to 3 moles P_i/mole protein using the commercial enzyme.

In contrast to the results of Ross and Bhatnagar (1989) and Ross (1989), Seguro and Motoki (1989) only found radiolabeled P_i incorporated into acidic polypeptides of glycinin and some uncharacterized low molecular weight fractions of soy protein. Basic polypeptides of glycinin, β-conglycinin and whey proteins were left unmodified. These authors also reported using a commercial enzyme, and under optimum phosphorylation conditions, only 2.0 moles P_i/mole acidic polypeptide were incorporated. Seguro and Motoki (1990) found that modified acidic polypeptides were less sensitive to Ca^{++} than nonphosphorylated material, and the emulsion properties were improved. Seguro, Nio, and Motoki (1986) developed a patented process using cAMPdPK for protein phosphorylation. Several food proteins, including soy protein, albumin, and casein, were reported to be phosphorylated in the patent, but the extent of phosphorylation was not given.

Food Protein Phosphorylation by Casein Kinase

Casein kinases exist as two distinct sets of enzymes (Pinna, Meggio, and Marchiori 1990). The casein kinase in mammary tissue that normally provides in vivo phosphorylation of nacent proteins is distinct from the multi-substrate and ubiquitous casein kinase (CK-2) that is responsible for in vitro casein phosphorylation. The substrate specificity of the two enzymes is also different. For the casein kinase from mammary gland, the recognition sequence corresponds to the tripeptide Ser/Thr-X-Glu/Ser-P or Ser/Thr-X-Glu/Ser in nonphosphorylated proteins, where X is any amino acid (Mercier, Grosclaude, and Ribadeau Dumas 1972). For CK-2, the recognition sites have been identified as Ser-Glu-Ala-Glu-Glu-Glu and Ser-Ala-Ala-Glu-Glu-Glu (Pinna, Meggio, and Marchiori 1990).

Besides its in vivo activity, mammary gland casein kinase can phosphorylate food proteins other than casein, as demonstrated by the in vitro phosphorylation of several food proteins as shown in Table 2-3 (Bingham and Farrell 1974). There is a marked preference of the enzyme for dephosphorylated caseins over native caseins, and for native caseins over other food proteins.

The ability of casein kinases to phosphorylate vegetable protein, especially soy protein, has not, to our knowledge, been investigated for either mammary gland kinase or CK-2. As noted previously, these kinases respond to specific amino acid sequences that contain the serine or threonine to be

TABLE 2–3   Phosphorylation of Proteins by a Casein Kinase from Rat Mammary Gland

Protein Substrate	Rate of Phosphate Addition (nmol/mg/20 min)
α_{s1}-Casein	8.6
Dephosphorylated α_{s1}-casein	43.0
β-Casein	2.5
Dephosphorylated β-casein	32.4
κ-Casein	1.3
Dephosphorylated κ-casein	8.5
β-lactoglobulin	2.1
α-lactalbumin	1.1
Fat globule membrane proteins	6.1
Histone-arginine rich	0.1
Histone-lysine rich	0.1
Phosvitin	0.3
Lysozyme	0.0

Source: From Bingham and Farrell 1974.

modified. An examination of the known primary structural sequences of glycinin and β-conglycinin revealed several potential phosphorylation sites for the mammary gland kinase in both glycinin and β-conglycinin, but no potential phosphorylation sites were observed for CK-2 (unpublished observations). Based on this information, mammary gland protein kinase should be evaluated for possible phosphorylation of soy protein. However, the type of information reported in Table 2–3 must be obtained on the affinities of casein kinase toward vegetable protein substrates, including soy, before serious consideration can be given to using this enzyme in protein modification on an industrial scale.

A description of the enzymatic phosphorylation of food proteins would be incomplete without noting a major hurdle to be overcome when contemplating scaleup of the modification. The phosphate donor, ATP, is the weak link in the phosphorylation process because of its relatively high cost. A carefully designed process must include provisions for the continuous generation of ATP. In this regard, Langer et al. (1976) suggested that acetate kinase (AK) was a good choice for a phosphotransferase to carry out the conversion of ATP from ADP after consideration of enzyme activity, stability, and cost. The substrate for this enzyme, acetyl phosphate, should be readily available, since Lewis, Haynie, and Whitesides (1979) have developed a method of producing this compound at low cost. The reactions catalyzed by AK and cAMPdPK could then be coupled in the following manner:

$$\text{acetyl phosphate} + \text{ADP} \underset{}{\overset{\text{AK}}{\rightleftharpoons}} \text{acetate} + \text{ATP}$$

$$\text{protein} + \text{ATP} \underset{}{\overset{\text{cAMPdPK}}{\rightleftharpoons}} \text{protein} - \text{P} + \text{ADP}$$

One possible scenario envisions the preceding reactions taking place in a hollow fiber reactor system similar to the one developed by Ishikawa et al. (1989) for the continuous production of glucose 6-phosphate from glucose. They used AK coupled with glucokinase to produce glucose 6-phosphate. In our example, the protein to be phosphorylated would take the place of glucose. The reactor system would be fed acetyl phosphate continuously and the protein would be circulated in a closed loop until phosphorylation was complete. The residual acetate would be separated from the phosphorylated protein at the end of the process.

CONCLUSIONS

We have described several chemical and enzymatic methods for food protein deamidation or phosphorylation. The methods discussed are relatively new. However, only a few show promise as food protein modifications that can be scaled up to meet FDA and food industry standards. We hope this review has pointed out both the opportunities and the pitfalls of developing a protein deamidation or phosphorylation process for the food industry. The road ahead will be rough, but given the ingenuity of scientists working to fully develop these modifications and what they have accomplished so far, success will only be a matter of time.

References

Adler-Nissen, J. 1986. *Enzymic Hydrolysis of Food Proteins,* pp. 324–329. New York: Elsevier Applied Science Publishers.

Aswad, D. W. 1984. Stoichiometric methylation of porcine adrenocorticotropin by protein carboxyl methyltransferase requires deamidation of asparagine 25. *J. Biol. Chem.* **259**:10,714–10,721.

Battersby, R., and J. C. Robinson. 1955. Studies on specific chemical fission of peptide links. I. Rearrangements of aspartyl and glutamyl peptides. *J. Chem. Soc.* **1955**:259–269.

Bercovici, D., H. F. Gaertner, and A. J. Puigserver. 1987. Transglutaminase-catalyzed incorporation of lysine oligomers into casein. *J. Agric. Food Chem.* **35**:301–304.

Bhatt, N. P., K. Patel, and R. T. Borchardt. 1990. Chemical pathways of peptide degradation. I. Deamidation of adrenocorticotropic hormone. *Pharm. Res.* **7**(6):593–599.

Bingham, E. W., and H. M. Farrell, Jr. 1974. Casein kinase from the golgi apparatus of lactating mammary gland. *J. Biol. Chem.* **249**(11):3647–3651.

Bjarnason, J., and K. J. Carpenter. 1969. Mechanisms of heat damage in proteins. I. Models with acylated lysine units. *Brit. J. Nutr.* **23**(4):859–868.

Brinegar, A. C., and E. Kinsella. 1980. Reversible modification of lysine in soybean proteins, using citraconic anhydride: Characterization of physical and chemical changes in soy protein isolate, the 7S globulin, and lypoxygenase. *J. Agric. Food Chem.* **28**:818–824.

Cohen, P. 1982. The role of protein phosphorylation in neutral hormonal control of cellular activity. *Nature* **296**:613–620.

Ellinger, R. H. 1972. Phosphates in food processing. In *CRC Handbook of Food Additives,* 2d ed., vol. 1, ed. T. E. Furia, p. 640. Cleveland, OH: CRC Press.

Feeney, R. E. 1977. Chemical modification of food proteins. In *Food Proteins,* ed. R. E. Feeney, and J. R. Whitaker, pp. 3–36. Washington DC: American Chemical Society.

Feeney, R. E., and J. R. Whitaker. 1985. Chemical and enzymatic modification of plant proteins. In *New Protein Foods,* vol. 5, ed. A. M. Altschul and H. L. Wilcke, pp. 181–219. New York: Academic Press.

Feeney, R. E., B. Yamasaki, and K. F. Geoghegan. 1982. Chemical modification of proteins: An overview. In *Modification of Proteins: Food, Nutritional, and Pharmacological Aspects,* ed. R. E. Feeney, and J. R. Whitaker, pp. 3–55. Washington, DC: American Chemical Society.

Feller, K. 1989. Characterization of a protein kinase from soybean seedlings. Study of the variation of calcium-regulated kinase activity during infection with the incompatible race 1 and the compatible race 3 of *Phytophthora Megasperma F. Sp. Glycinea* by *in vitro* phosphorylation of calf thymus histone H1. *Plant Sci.* **60**:67–75.

Folk, J. E., and J. S. Finlayson. 1977. The ε-(γ-glutamyl)-lysine crosslink and the catalytic role of transglutaminases. In *Advances in Protein Chemistry,* vol. 31, ed. C. B. Anfinsen, J. T. Edsall, and F. M. Richards, p. 4. New York: Academic Press.

Geiger, T., and S. Clarke. 1987. Deamidation, isomerization, and racemization at asparaginyl and aspartyl residues in peptides. *J. Biol. Chem.* **262**:785–794.

Gill, B. P., A. J. O'Shaughnessey, P. Henderson, and D. R. Headon. 1985. An assessment of potential of peptidoglutaminase I and II in modifying the charge characteristics of casein and whey proteins. *Irish J. Food Sci. Technol.* **9**:33–41.

Gowda, S., and D. T. N. Phillay. 1982. Cyclic AMP independent protein kinases from soybean cotyledons. *Plant Sci. Lett.* **25**:49–59.

Hamada, J. S. 1989. "Potential of Gel Adsorption and Ultrafiltration for Immobilization and Multiuse of Peptidoglutaminase." Paper read at 50th Annual Meeting of the Institute of Food Technology, 25–29 June 1989, Chicago, IL.

Hamada, J. S. 1990a. "Peptidoglutaminase Deamidation of Proteins for Improved Food Use." Paper read at 81st Annual Meeting of the American Oil Chemists' Society, 22–25 April 1990, Baltimore, MD.

Hamada, J. S. 1990b. "A Batch Ultrafiltration Reactor for Large-scale Peptidoglutaminase Deamidation of Food Proteins." Paper read at 51st Annual Meeting of the Institute of Food Technology, 16–20 June 1990, Anahiem, CA.

Hamada, J. S., and W. E. Marshall. 1988. Enhancement of peptidoglutaminase deamidation of soy proteins by heat treatment and/or proteolysis. *J. Food Sci.* **53**:1132–1134, 1149.

Hamada, J. S., and W. E. Marshall. 1989. Preparation and functional properties of enzymatically deamidated soy proteins. *J. Food Sci.* **54**:598–601, 635.

Hamada, J. S., F. F. Shih, A. W. Frank, and W. E. Marshall. 1988. Deamidation of soy peptides and proteins by *Bacillus circulans* peptidoglutaminase. *J. Food Sci.* **53**:671–672.

Hanks, S. K., A. M. Quinn, and T. Hunter. 1988. The protein kinase family: Conserved features and deduced phylogeny of the catalytic domains. *Science* **241**:42–52.

Harding, J. J. 1985. Nonenzymatic covalent posttranslational modification of protein *in vivo*. *Adv. Protein Chem.* **37**:247–334.

Hirotsuka, M., H. Taniguchi, H. Narita, and M. Kito. 1984. Functionality and digestibility of a highly phosphorylated soybean protein. *Agric. Biol. Chem.* **48**:93–100.

Huang, Y. T., and J. E. Kinsella. 1987. Effects of phosphorylation on emulsifying and foaming properties and digestibility of yeast protein. *J. Food Sci.* **52**(6):1684–1688.

Hunter T., and J. A. Cooper. 1985. Protein-tyrosine kinases. *Ann. Rev. Biochem.* **54**:897–930.

Ishikawa, H., S. Takase, T. Tanaka, and H. Hikita. 1989. Experimental investigation of G6P production and simultaneous ATP regeneration by conjugated enzymes in an ultrafiltration hollow-fiber reactor. *Biotechnol. Bioeng.* **34**:369–379.

Kato, A., Y. Lee, and K. Kobayashi. 1989. Deamidation and functional properties of food proteins by the treatment with immobilized chymotrypsin at alkaline pH. *J. Food Sci.* **54**(5):1345–1347, 1372.

Kato, A., A. Tanaka, Y. Lee, N. Matsudomi, and K. Kobayashi. 1987*a*. Effects of deamidation with chymotrypsin at pH 10 on the functional properties of proteins. *J. Agric. Food Chem.* **35**:285–288.

Kato, A., A. Tanaka, N. Matsudomi, and K. Kobayashi. 1987*b*. Deamidation of food proteins by protease in alkaline pH. *J. Agric. Food Chem.* **35**:224–227.

Kikuchi, M., H. Hayashida, E. Nakano, and K. Sakahuchi. 1971. Peptidoglutaminase: Enzymes for selective deamidation of γ-amide of peptide-bound glutamine. *Biochemistry* **10**(7):1222–1229.

Kossiakoff, A. A. 1988. Tertiary structure is a principal determinant to protein deamidation. *Science* **240**:191–194.

Krebs, E. G. 1986. The enzymology of control by phosphorylation. In *The Enzymes,* vol. 17, ed. P. D. Boyer and E. G. Krebs, pp. 3–20. New York: Academic Press.

Krebs, E. G., and J. A. Beavo. 1979. Phosophorylation-dephosphorylation of enzymes. *Ann. Rev. Biochem.* **48**:923–959.

Langan, T. A. 1973. Protein kinases and protein kinase substrates. *Adv. Cyclic Nucleotide Res.* **3**:99–153.

Langer, R. S., B. K. Hamilton, C. R. Gardner, M. C. Archer, and C. K. Colton. 1976. Enzymatic regeneration of ATP. I. Alternative routes. *AIChE J.* **22**(6):1079–1090.

Lewis, J. M., S. L. Haynie and G. M. Whitesides. 1979. An improved synthesis of diammonium acetyl phosphate. *J. Org. Chem.* **44**(5):864–865.

Lin, P. P., and J. L. Key. 1980. Histone kinase from soybean hypocotyls. *Plant Physiol.* **66**:360–367.

Lorand, L., and S. Conrad. 1984. Transglutaminases. *Mol. Cell. Biochem.* **58**:9–35.

Lura, R., and V. Schirch. 1988. Role of peptide conformation in the rate and mechanism of deamidation of asparaginyl residues. *Biochemistry* **27**:7671–7677.

Matheis, G., and J. R. Whitaker. 1984. Chemical phosphorylation of food proteins: An overview and prospectus. *J. Agric. Food Chem.* **32**:699–705.

Matheis, G., M. H. Penner, R. E. Feeney, and J. R. Whitaker. 1983. Phosphorylation of casein and lysozyme by phosphorus oxychloride. *J. Agric. Food Chem.* **31**(2):379–387.

Matsudomi, N., T. Sasaki, A. Kato, and K. Kobayashi. 1985. Conformational changes and functional properties of acid-modified soy protein. *Agric. Biol. Chem.* **49**(5):1251–1256.

Meinwald, Y. C., E. R. Stinson, and A. Scheraga. 1986. Deamidation of the asparaginyl-glycyl sequence. *Int. J. Peptide Protein Res.* **28**:79–84.

Mercier, J.-C., F. Grosclaude, and B. Ribadeau Dumas. 1972. Primary structure of bovine caseins. *Milchwissenschaft* **27**:402–408.

Meyer, E. W., and L. D. Williams. 1977. Chemical modification of soy proteins. In *Food Proteins* ed. R. E. Feeney and J. R. Whitaker, pp. 52–66. Washington DC: American Chemical Society.

Motoki, M., K. Seguro, N. Nio, and K. Takinami. 1986. Glutamine-specific deamidation of α_{s1}-casein by transglutaminase. *Agric. Biol. Chem.* **50**(12):3025–3030.

Murray, M. G., T. J. Guilfoyle, and J. L. Key. 1978. Isolation and characterization of a chromatin-associated protein kinase from soybean. *Plant Physiol.* **61**:1023–1030.

Mycek, M. J. and H. Waelsch. 1960. The enzymatic deamidation of proteins. *J. Biol. Chem.* **235**(12):3513–3517.

Nestler, E. J., S. I. Walaas, and P. Greengard. 1984. Neuronal phosphoproteins: Physical and clinical implications. *Science* **225**:1357–1364.

Okuno, S., and H. Fujisawa. 1990. Stabilization, purification and crystalization of catalytic subunit of cAMP-dependent protein kinase from bovine heart. *Biochim. Biophys. Acta* **1038**(2):204–208.

Patel, K., and R. T. Borchardt. 1990*a*. Chemical pathways of peptide degradation. II. Kinetics of deamidation of an asparaginyl residue in a model hexapeptide. *Pharm. Res.* **7**(7):703–711.

Patel, K., and R. T. Borchardt. 1990*b*. Chemical pathways of peptide degradation. III. Effect of primary sequence on the pathways of deamidation of asparaginyl residues in hexapeptides. *Pharm. Res.* **7**(8):787–793.

Pinna, L. A., F. Meggio, and F. Merchiori. 1990. Type-2 casein kinases: General properties and substrate specificity. In *Peptides and Protein Phosphorylation,* ed. B. C. Kemp, pp. 145–169. Boca Raton, FL: CRC Press.

Polya, G. M., and J. R. Davies. 1982. Resolution of Ca^{2+}-calmodulin-activated protein kinase from wheat germ. *FEBS Lett.* **150**:167–171.

Putnam-Evans, C. L., A. C. Harmon, and M. J. Cormier. 1990. Purification and characterization of a novel calcium-dependent protein kinase from soybean. *Biochemistry* **29**:2488–2495.

Robinson, A. B., J. W. Scotchler, and J. H. McKerrow. 1973. Rates of nonenzymatic deamidation of glutaminyl and asparaginyl residues in pentapeptides. *J. Am. Chem. Soc.* **95**:8156–8189.

Ross, L. F. 1989. Optimization of enzymatic phosphorylation of soybean storage proteins: Glycinin and β-conglycinin. *J. Agric. Food Chem.* **37**(5):1257–1261.

Ross, L. F., and D. Bhatnagar. 1989. Enzymatic phosphorylation of soybean proteins. *J. Agric. Food Chem.* **37**(4):841–844.

Rubin, C. S., and O. M. Rosen. 1975. Protein phosphorylation. *Ann. Rev. Biochem.* **44**:831–887.

Seguro, K., and M. Motoki. 1989. Enzymatic phosphorylation of soybean proteins by protein kinase. *Agric. Biol. Chem.* **53**(12):3263–3268.

Seguro, K., and M. Motoki. 1990. Functional properties of enzymatically phosphorylated soybean proteins. *Agric. Biol. Chem.* **54**(5);1271–1274.

Seguro, K., S. Nio, and M. Motoki. 1986. The Manufacture method of Modified Proteins. Japanese Patent No. 128,843 (March, 1986).

Shih, F. F. 1987. Deamidation of protein in a soy extract by ion exchange resin catalysis. *J. Food Sci.* **52**(6):1529–1531.

Shih, F. F. 1989. Partially Deamidated Oilseed Proteins and Process for the Preparation Thereof. U.S. Patent 4,824,940 (Apr. 25, 1989).

Shih, F. F. 1990a. Deamidation during treatment of soy protein with protease. *J. Food Sci.* **55**(1):127–129, 132.

Shih, F. F. 1990b. Deamidation studies on selected food proteins. *J. Am. Oil Chem. Soc.* **67**(10):675–677.

Shih, F. F. 1991. Effect of anions on the deamidation of soy protein *J. Food Sci.* **56**(2):452–454.

Shih, F. F., and A. D. Kalmar. 1987. SDS-catalyzed deamidation of oilseed proteins. *J. Agric. Food Chem.* **35**(5):672–675.

Shoji, S., D. C. Parmelee, R. D. Wade, S. Kumar, L. H. Ericsson, and K. A. Walsh. 1981. Complete amino acid sequence of the catalytic subunit of bovine muscle cyclic AMP-dependent protein kinase. *Proc. Natl. Acad. Sci. (USA)* **78**:848–851.

Shoji, S., L. H. Ericsson, K. A. Walsh, E. H. Fischer, and K. Tetani. 1983. Amino acid sequence of the catalytic subunit of bovine type II adenosine cyclic 3′,5′-phosphate-dependent protein kinase. *Biochemistry* **22**:3702–3709.

Smith, S. B., J. B. White, J. B. Siegel, and E. G. Krebs. 1981. Cyclic AMP-dependent protein kinase: Primary steps of allosteric regulation. In *Protein Phosphorylation,* ed. O. R. Rosen and E. G. Krebs, pp. 55–65. Cold Spring Harbor, ME: Cold Spring Harbor Laboratory.

Sondheimer, E., and R. W. Holley. 1954. Imides form asparagine and glutamine. *J. Am. Chem. Soc.* **76**:2467–2470.

Sung, H., H. Chen, T. Liu, and J. Su. 1983. Improvement of the functionalities of soy protein isolate through chemical phosphorylation. *J. Food Sci.* **48**:716–721.

Uhler, M. D., and G. S. McKnight. 1987. Expression of cDNAs for two isoforms of the catalytic subunit of cAMP-dependent protein kinase. *J. Biol. Chem.* **262**:15,202–15,207.

Uhler, M. D., J. C. Chrivia, and G. S. McKnight. 1986. Evidence for a second isoform of the catalytic subunit of cAMP-dependent protein kinase. *J. Biol. Chem.* **261**:15,360–15,363.

Uhler, M. D., D. F. Carmichael, D. C. Lee, J. C. Chrivia, E. G. Krebs, and G. S. McKnight. 1986. Isolation of the cDNA clones coding for the catalytic subunit of mouse cAMP-dependent protein kinase. *Proc. Natl. Acad. Sci. (USA)* **83**:1300–1304.

Whitaker, J. R. 1977. Enzymatic modification of proteins applicable to foods. In *Food Proteins,* ed. R. E. Feeney and J. R. Whitaker, pp. 95–155. Washington DC: American Chemical Society.

Whitaker, J. R., and A. J. Puigserver. 1982. Fundamentals and applications of enzymatic modifications of proteins: An overview. In *Modification of Proteins: Food, Nutritional, and Pharmacological Aspects,* ed. R. E. Feeney, and J. R. Whitaker, pp. 57–87. Washington DC: American Chemical Society.

Woo, S. L., and T. Richardson, 1983. Functional properties of phosphorylated β-lectoglobulin. *J. Dairy Sci.* **66**:984–987.

Woo, S. L., L. K. Creamer, and T. Richardson, 1982. Chemical phosphorylation of bovine β-lactoglobulin. *J. Agric. Food Chem.* **30**:65–70.

3

Natural Enzyme and Biocontrol Methods for Improving Fruits and Fruit Quality

Elizabeth A. Baldwin and Robert A. Baker

The fresh fruit industry has many postharvest problems to overcome. Often fruits, such as tomato or banana, are harvested immature in order to survive shipping and handling conditions. During transit and upon arrival at retail markets, some fruits are either stored at low temperatures or in controlled atmospheres to delay ripening in order to extend shelf life. Other fruits are treated with ethylene gas to accelerate ripening prior to display in supermarkets. Such practices are costly and can result in a poor-quality product.

Current basic research is aimed at understanding fruit ripening, with the goal of finding new ways to manipulate this process to extend shelf life and improve fruit quality. Major commercial benefits could be gained from the identification of molecular and genetic control mechanisms in fruit ripening. The following sections discuss the use of enzymes, microorganisms, and other molecular signals that have the potential to influence ripening and otherwise improve the quality of stored fresh or processed fruits.

Climacteric fruits, such as tomato, produce relatively high levels of ethylene gas and carbon dioxide upon ripening, while nonclimacteric fruits, such as citrus, ripen more slowly without this phenomenon (Biale and Young 1981). Both of these fruits have been studied extensively as examples of the two types of ripening systems based on their respective respiratory patterns. In both ripening scenarios, sugars, acids, texture, pigments, and flavor volatiles play important roles in the final quality of the ripened product. In

---

climacteric fruit particularly, the increased production of ethylene promotes many aspects of ripening. Manipulation of ethylene production, therefore, results in the acceleration or delay of many ripening events, such as softening and color development (Wills et al. 1981). Increasing ethylene production generally promotes ripening and senescence of fruits and vegetables, especially of those in the climacteric category. On the other hand, inhibition of ethylene production usually retards ripening and senescence in such fruit. Certain molecular signals, which can be used to induce ethylene production and thus promote ripening, are discussed later in this chapter. Enzymes that digest cell walls are available from both plant and fungal sources. These enzymes have been shown to elicit ethylene production and can affect the texture of fruits and vegetables. Some data will be included where these enzymes were used to initiate ethylene production and improve the quality of processed fruit products.

Resistance of stored fruits to pathogens is also related to shelf life and quality. In breeding and selecting for desirable horticultural characteristics, such as low tannin, thin skins, and high sugar, we may be developing varieties that are more susceptible to postharvest diseases (Wilson and Wisniewski 1989). Fungicides are commonly used to combat decay, but they often result in resistant strains of phytopathogens (Jeffries and Jeger 1990) and have come under fire for presenting potential risks to health and environment. The use of natural agents to control postharvest diseases holds possibilities for the replacement of unpopular chemicals. The potential benefits of "biocontrol" for the fresh fruit industry are presented toward the end of the chapter.

FRUIT RIPENING

Importance of Ethylene

Ethylene is a gaseous plant hormone that regulates many plant responses and promotes ripening in climacteric fruit such as tomato. Ethylene also influences color development in some nonclimacteric fruits such as citrus. The latter effect may be related more to senescence than ripening (Wills et al. 1981). A sharp increase in CO_2 and ethylene production is exhibited during tomato fruit ripening. This gas production is accompanied by flavor volatile synthesis, softening, chlorophyll breakdown, and lycopene development (Khudairi 1972; Grierson 1985; Baldwin, Nisperos-Carriedo, and Moshonas 1991). Initiation and continuance of texture and observed color changes are thought to be controlled to some extent by the elevated levels of ethylene (Mizrahi, Dostal, and Cherry 1975; Jeffery et al. 1984). In mature green tomatoes, ethylene stimulates synthesis of the red pigment lycopene and endopolygalacturonase (endo-PG, EC 3.2.1.15), an enzyme that may affect fruit texture (Grierson and Tucker 1983).

Importance of Cell-Wall-Modifying Enzymes

Endopolygalacturonase

Many fruits soften as they ripen. In tomato, peach, pear, and papaya endopolygalacturonase (endo-PG) is associated with cell-wall changes, ripening, and softening (Pressey 1986). α-1,4-Galacturonide glycanohydrolase, or endo-PG, is one of several classes of pectinases that randomly cleave galacturonan chains in pectin, a major component of the middle lamella in plant cell walls (Table 3-1). This enzyme is a depolymerase, catalyzing the hydrolytic cleavage of α-1.4-glycosidic bonds in the D-galacturonan moiety of pectic substances, and is specific for de-esterified galacturonans, that is, polypectate. PGs are produced by both microorganisms and higher plants (Rexova-Benkova and Markovic 1976). The products of endo-PG action on isolated cell walls are pectic fragments of various lengths.

Pectinmethylesterase

Pectinmethylesterase (PME) is a carboxyl ester hydrolase (3.1.1.11) that de-esterifies the methyl esters of pectin (Table 3-1), giving rise to blocks of free carboxyl groups (Rexova-Benkova and Markovic 1976). It is suggested that at least two adjacent free carboxyl groups are necessary for endo-PG action to occur. The requirement for de-esterification by endo-PG suggests a dependency of this enzyme on PME, although evidence has not been provided to support this supposition. Both tomatoes and citrus contain high levels of PME, but only tomatoes also contain endo-PG. This implies that PME

TABLE 3-1   Nomenclature of Some Pectin-degrading Enzymes

Name Classification	Abbre- viation	EC Number	Substrate	Product

Endopolygalacturonase
  hydrolase/random
	endo-PG	EC 3.2.1.15	D-Galacturonan
			  (polypectate)
			  (pectic acid)
				Simple oligou-
				  ronides

Exopolygalacturonase
  hydrolase/terminal
	exo-PG	EC 3.2.1.67	D-Galacturonan	Galacturonic
				  acid

Pectinmethylesterase
  carboxyl ester hydro-
  lase
	PME	EC 3.1.1.11	Polymethyl es-
			  ters of D-
			  galacturonans
				De-esterified
				  pectin

Endopectin lyase
  transeliminase/
  random
	endo-PL	EC 4.2.2.10	Esterified D-
			  galacturonan
			  (pectin)
				Esterified un-
				  saturated oli-
				  gouronides

must have another role in cell-wall modification other than substrate preparation for cell-wall hydrolysis by endo-PG.

Exopolygalacturonase

Exopolygalacturonase (exo-PG, EC 3.2.1.67) also occurs in tomato as well as in other fruits and plant tissues (Pressey and Avants 1977; Pressey 1987). This enzyme removes galacturonic acid monomer units from the nonreducing ends of pectin chains (Table 3–1). Galacturonic acid is therefore the resulting product of exo-PG action on plant cell walls.

Relationship between Cell-Wall Changes and Ethylene

Synthesis of both ethylene and polygalacturonase increases considerably during ripening of tomato and other fruits (Pressey 1986). It has been reported that the increased levels of ethylene, produced during ripening, promoted *de novo* synthesis of polygalacturonase as well as lycopene in tomato fruit (Grierson and Tucker 1983). Meanwhile, cell-wall fragments, produced by fungal pectinases, elicited ethylene production in pear cell cultures (Tong, Labavitch, and Yang 1986). Albersheim and others have noted that both fungal and plant cell-wall fragments elicited phytoalexin production (thought to be a plant defense response) in several plant species. They have suggested that such fragments may control various physiological responses of plants, such as rate of cell growth, time of flowering, and other activation mechanisms for resistance to potential pathogens (McNeil et al. 1984). Galacturonic acid, the product of exo-PG action on cell walls, has also been reported to induce ethylene production when infiltrated into mature green tomato fruit (Kim, Gross, and Solomos 1987). This raised the question whether endogenous tomato endo-PG and/or exo-PG could be contributing to the high levels of ethylene production observed in ripening fruit via release of cell-wall products such as pectin oligomers or galacturonic acid.

USE OF PECTINASES AND THEIR CELL-WALL SUGAR PRODUCTS TO RIPEN TOMATOES

Ripening Tomatoes with Endogenous Polygalacturonase Enzymes

Endopolygalacturonase

Endo-PG and PME were partially purified from 'Better Boy' tomatoes. These enzymes were then vacuum-infiltrated together or separately through the stem scar into mature green tomatoes of the small fruit variety (cherry tomatoes) that were not yet synthesizing these enzymes or elevated levels of ethylene. Infiltration of endo-PG alone in a buffer of pH 4.5, which is optimal for activity, increased ethylene production by four- to twelvefold compared to the two buffer controls and accelerated ripening (Baldwin and Pressey 1988a). The effect of this enzyme on tomato ripening was greatly reduced when the buffer pH was raised to 7.0, where the enzyme is not very active. This would suggest that enzyme activity rather than physical presence of the protein was required for the ethylene response. In some cases addition of PME to the endo-PG treatment enhanced the ethylene response, perhaps due to demethylation of pectin. This would provide endo-PG with the preferred substrate of polypectate. Endo-PG and PME also induced ethylene but not ripening tomato mutants *rin* and *nor.*

Exopolygalacturonase

In subsequent studies it was discovered that exo-PG from several plant sources (including tomato) also elicited ethylene production and accelerated ripening when infiltrated into mature green tomato fruit (Baldwin and Pressey 1990). Infusion of fruit with exo-PG was not as effective, however, in terms of accelerated ripening as the endo-PG treatments.

Ripening Tomatoes with Polygalacturonase Hydrolysis Products

Products of Endopolygalacturonase

Ethylene production and ripening was stimulated by infiltrating tomato pectin fragments into mature green tomatoes (Brecht and Huber 1988). These pectin fragments resulted from autohydrolysis of tomato cell walls in vitro, presumably due to the presence of endo-PG. The resulting pectin fragments were separated from neutral-cell-wall sugars by ion exchange chromatography and further separated by size using gel filtration. Significant ethylene production increases were observed, with tomato pectin oligomers having a degree of polymerization (DP) greater than eight (more than eight galacturonic acid units). This study supports the hypothesis that endo-PG elicitation of ethylene and acceleration of ripening may have been due to enzyme-produced pectic fragments. Another study showed that citrus pectin oligomers (DP 2 − 10 and > 10, released from citrus pectin by a commercial pectinase) also induced mature green cherry tomatoes to produce ethylene (Baldwin and Pressey 1988b).

Products of Exopolygalacturonase

Galacturonic acid, as mentioned previously, is the product of exo-PG action on plant cell walls. Mature green tomatoes, infiltrated with galacturonic acid, showed an increased ethylene production pattern similar to that observed with exo-PG (Kim, Gross, and Solomos 1987; Baldwin and Pressey 1990). Other sugars, such as galactose and mannose, also elicited an ethylene response, while fructose, glucose, lactose, raffinose, rhamnose, sorbitol, sucrose, and xylose did not (Kim, Gross, and Solomos 1987). This presents the possibility that the exo-PG elicitation of ethylene may have been due to the release of galacturonic acid from pectin in cell walls.

Ripening Tomatoes with Fungal Pectinase Isoenzymes

Pectolyase is a commercially available enzyme mixture prepared from culture filtrates of *Aspergillus japonicus* (Ishii 1976). The enzymes contained in this mixture are very effective in degrading cell walls and thereby liberating protoplasts. Having observed that purified tomato PG-induced ethylene production and accelerated ripening when infiltrated into green tomato fruit, these studies were expanded to include microbial pectic enzymes, and the commercial enzyme mixture, Pectolyase, was selected as a source. Attempts to separate the pectinases in Pectolyase revealed that this organism produces several isoenzymes of two types of pectin-digesting enzymes (Baldwin and Pressey 1989). Two isoenzymes of endopolygalacturonase and three of endopectin lyase (Endo-PL, EC 4.2.2.10), as well as a cellulase and a pectinmethylesterase, were detected and partially purified by anion exchange chromatography.

Infiltration of the Pectolyase mixture into mature green tomato fruit revealed that these enzymes were also effective in inducing ethylene production in tomatoes as well as color development (Baldwin and Pressey 1989). This enzyme preparation elicited ethylene production in the nonripening mutant tomatoes, *rin* and *nor,* as well, but not color development.

Endopolygalacturonase Isoenzymes

The two purified endo-PG isoenzymes from Pectolyase were adjusted to equal activities and individually infiltrated into mature green tomato fruit (Table 3–2). Surprisingly, one isoenzyme was much more effective than the other in inducing ethylene and subsequent ripening. The more efficient enzyme in inducing ethylene production was also more efficient in solubilizing uronic acids from pectin polymers in plant cell walls (Baldwin and Pressey 1989). This would lead to the release of pectin oligomers, possible elicitors of ethylene production.

Endo-Pectin Lyase Isoenzymes

Pectin lyases or *trans*-eliminases cleave the α-1,4-galacturonosidic bond by a *trans*-elimination reaction, producing an unsaturated bond in the galac-

TABLE 3-2   Effects of Pectolyase Enzymes on Ethylene Production
in Enzyme-Infiltrated Tomato Fruit

Hours after Infiltration	PL-A	PL-C	PG-A	PG-B	Buffer
	(nl Ethylene/Fruit/hr)				
1	8 ± 4	9 ± 2	1 ± 1	1 ± 1	0 ± 0
3	41 ± 5	44 ± 7	41 ± 8	37 ± 7	24 ± 11
6	31 ± 16	129 ± 12	92 ± 21	53 ± 20	8 ± 5
10	70 ± 32	48 ± 15	9 ± 9	1 ± 1	2 ± 1
19	76 ± 15	18 ± 7	3 ± 1	1 ± 1	2 ± 1
24	55 ± 7	11 ± 8	3 ± 1	2 ± 1	2 ± 1

Source: After Baldwin and Pressey 1989.

Note: PL-A and PL-B are isoenzymes of endo-PL, while PG-A and PG-B are isoenzymes of endo-PG that had been separated from the commercial enzyme mixture, Pectolyase. Volume of enzyme solutions corresponding to 1% of tomato weight and containing 83 units/ml of PL or 122 units/ml of PG were vacuum infiltrated into mature green cherry tomatoes. The fruit were periodically sealed in glass jars for ethylene determinations. Data are means of four replications ± SE.

turonic acid residue at the nonreducing end of the fragment released (Table 3-1). The enzymes from this study were specific for esterified galacturonans. Lyases are common in microorganisms but not in plants (Pressey 1986). The purified endo-PL isoenzymes were much more effective than the endo-PGs in degrading cell walls (Baldwin and Pressey 1989). The relative ineffectiveness of the PGs may have been due to highly esterified pectin in the cell walls, which is not the preferred substrate of these enzymes. The PGs would react optimally in the presence of PME that had been removed by ion chromatography. In contrast, had the PME been added back to the reaction mixture, it would have reduced the reactivity of the lyases by de-esterifying the pectin in the cell walls. In any event, the PLs were also generally more efficient at eliciting ethylene and accelerating ripening in mature green tomato fruit, presumably due to their effective release of pectin fragments (Table 3-2). It is also possible that the unsaturated pectin fragments (oligouronides) released by pectin lyases are more effective than the simple oligouronides released by PGs.

Whether the ethylene response to enzyme infiltration is a ripening response to a ripening signal or a wound response to a wound signal is as yet unclear, but either response can initiate ethylene production and color change. Improved methods of infiltration are needed, however, before this process can be used to produce fruit of acceptable quality. Currently, the fruit exhibit blotchy ripening due to uneven infiltration of enzyme solutions.

USE OF FUNGAL PECTINASES
TO PROMOTE COLOR IN CITRUS

Response to Enzymes

Cell-wall changes and ethylene production are often correlated in physiological events, such as mechanical wounding (Evensen, Bausher, and Biggs 1981), abscission (Ketring and Melouk 1982), and pathogen invasion (DeLaat and VanLoon 1982) as well as in ripening of certain fruits. In all cases, cell-wall fragments are produced. Chlorophyll breakdown (green color) and carotenoid synthesis (orange color) can be promoted in oranges by treatment with exogenous ethylene gas. Oranges, which do not exhibit elevated levels of ethylene production and respiration during ripening, were induced to do so when injected with the Pectolyase enzyme mixture. This promoted color development in the peel proximal to the sites of enzyme injection (Baldwin and Biggs 1988). As with PG-treated tomato, optimal ethylene production was observed at the optimal pH range for the PG and PL isoenzymes.

Endopolygalacturonase

Mature oranges do not contain endogenous endo-PG as do tomatoes. However, when the endo-PG isoenzymes from Pectolyase were injected individually into oranges, both enzymes induced substantial ethylene production and color development after a considerable lag time (Baldwin and Pressey 1989). The enzyme-induced ethylene production lasted longer in oranges than the transient response observed in tomato (Table 3-3).

Endopectin Lyase

As in the tomato, the endo-PLs from Pectolyase were more effective than the endo-PGs in inducing ethylene production in citrus (Baldwin and Pressey 1989). The lag time for response to endo-PL treatment was much less than that of the endo-PG treatments (Table 3-3). The ethylene response occurred within a few hours of treatment, similar to what was observed with tomato. Again, the duration of the ethylene response to lyase enzyme treatment in citrus was considerably longer than the transient burst that occurred in tomato.

Response to Pectic Fragments

Acid-Digested Pectin

The oranges also exhibited elevated ethylene production when treated with acid-hydrolyzed pectin fragments, with the duration of the response lasting

TABLE 3–3.   Effects of Pectolyase Enzymes on Ethylene Production in Oranges

Hours after Infiltration	PL-A	PL-C	PG-A	PG-B	Buffer
		(nl Ethylene/Fruit/hr)			
1	38 ± 32	40 ± 17	0 ± 0	1 ± 2	1 ± 2
2	138 ± 47	49 ± 14	3 ± 2	16 ± 8	3 ± 4
5	232 ± 14	62 ± 6	21 ± 8	26 ± 20	3 ± 2
8	365 ± 164	119 ± 13	88 ± 7	168 ± 36	27 ± 8
18	318 ± 117	91 ± 34	122 ± 20	125 ± 32	9 ± 10
24	65 ± 46	93 ± 29	74 ± 18	104 ± 31	5 ± 8
42	26 ± 24	218 ± 151	47 ± 6	84 ± 26	6 ± 16
66	32 ± 10	57 ± 21	41 ± 11	80 ± 14	9 ± 16

Source: Data from Baldwin and Pressey 1989.

Note: PL-A and PL-C are isoenzymes of endo-PL, while PG-A and PG-B are isoenzymes of endo-PG that had been separated from the commercial enzyme mixture, Pectolyase. Enzyme solutions containing 68 units/ml of PL or 103 units/ml of PG were injected at six locations around the equator of the fruit (20 μl/ injection). The fruit were periodically sealed in glass jars for ethylene determinations. Data are means of four replications ± SE.

longer than observed for the Pectolyase mixture itself (Baldwin and Biggs 1988). When the acid-digested mixture was separated by size, certain size fragments showed greater ethylene-inducing capacity than others.

Enzyme-Digested Pectin and Polypectate

Pectolyase-digested pectic materials also induced ethylene production in citrus (Baldwin and Biggs 1988). Pectolyase-digested polypectate and citrus pectin both elicited ethylene, with the enzyme-digested citrus pectin being the more effective treatment. Enzyme digestion of polypectate (de-esterfied pectin) was assumed to be due to the action of endo-PG, since the endo-PL in Pectolyase does not attack pectic acid (Ishii and YoKotsuka 1975). Conversely, enzyme digestion of pectin (contains methyl groups) was thought to be primarily due to the action of endo-PL, since the endo-PG in Pectolyase is only 50% effective for digestion of methylated (esterified) pectin (Baldwin and Pressey 1989). Pectin fragment products produced by endo-PL contain a double bond, while fragments resulting from action of endo-PG do not (Table 3–1). The response to Pectolyase may therefore be due to the pectic fragments produced by the pectinases that in turn elicit ethylene production.

Enzyme digestion of cell walls in citrus does not induce the autocatalytic climacteric ethylene response as with tomato. The ethylene produced is more likely a senescence or wound response, as would occur in the case of pathogen invasion where similar enzymes are produced by fungal and bacterial pathogens in order to gain entrance into plant tissue. Nevertheless, the ethylene produced induced clorophyll breakdown and/or carotenoid synthesis, resulting in improved color from a commercial standpoint as well as abscission of the fruit from the stem. This is similar to methods currently used to degreen citrus with exogenous ethylene gas (Wills et al. 1981). Improved methods for application of commercial enzymes using vacuum infusion, dips, wax coatings, or sprays are under investigation.

USE OF PECTINMETHYLESTERASE
AND CALCIUM TO INCREASE FIRMNESS
IN PEACHES

Freestone peaches have equal levels of endo-PG and exo-PG, exhibit degradation of pectin, and soften markedly during ripening. Clingstone peaches, on the other hand, contain exo-PG but little or no endo-PG, exhibit less pectin degradation, and soften considerably less than their freestone counterparts (Pressey 1986). PME is postulated to increase firmness of fruits and vegetables by de-esterification of pectin, which allows for chelation of calcium. Peach halves were blanched to inactivate natural enzymes prior to vacuum infusion with PME extract (extracted from grapefruit pulp) containing 100 mg/l calcium chloride and subsequently canned (Javeri, Toledo, and Wicker 1990). Enzyme/calcium-treated peaches showed an increase in firmness up to four times that of nontreated controls.

The mechanism of increase in firmness of fruits and vegetables is not known. The effect of calcium on increasing firmness had been attributed to the formation of an eggbox structure through calcium crosslinks with carboxyl groups on de-esterified pectin. This does not seem to be the case in cucumber tissue, however, where McFeeters (1989) suggests that calcium inhibits softening by binding at sites other than pectin carboxyl groups. Indeed, in canned peaches the contribution of calcium to increasing firmness was secondary to the effect of PME activity.

USE OF FUNGAL NARINGINASE TO
REDUCE GRAPEFRUIT BITTERNESS

Effectiveness of Peel Infusion

Grapefruit peel and segment membranes, which may constitute 50% of the fruit by weight, are not usually considered edible. Thus this rich source of both soluble and insoluble food fiber is often discarded. A primary deterrent to consumption of grapefruit albedo is the very high content (as much as 2–5%) (Sinclair 1972) of an intensely bitter glucoside, naringin. Roe and

Bruemmer (1976) proposed treating albedo with naringinase to reduce the level of naringin, and increasing its palatability by inclusion of sweeteners, flavorings, and nutrients. This would permit a previously neglected food source to be utilized; in addition, the unusually porous nature of grapefruit albedo provides an ideal matrix for incorporation of other nutrients.

Optimum penetration of albedo with naringinase solutions was effected with vacuum infusion. To accomplish this, the flavedo (containing the oil glands) was removed with a commercial peeler, and the partially peeled fruit was submerged in a solution of fungal naringinase before infusion. Treatment with pectinase-free naringinase reduced the naringin level of mature 'Marsh' grapefruit albedo by 81%, a level adequate for a taste panel to determine as significantly less bitter. In further work it was found that early season (fall) grapefruit were more difficult to debitter, requiring approximately 10 times as much naringinase as more mature fruit (Roe and Bruemmer 1977). In addition, early season fruit contained other bitter components (limonin, coumarins, etc.) that were not reduced by naringinase treatment or masked by added sweeteners. Debittering of naringinase-treated grapefruit continued during storage at 4°C, such that taste panels preferred treated grapefruit stored for 33 to 46 days over freshly treated fruit.

Potential for Quality Improvement

Addition of flavored and sweetened gels (Jello) during the infusion process improved both palatability and appearance of the peeled fruit, and presence of a gelling agent prevented leakage of added liquids. The presence of added sweetener has been shown to increase the taste threshold of naringin, so that sweetened samples are perceived as less bitter than unsweetened samples (Guadagni, Maier, and Turnbaugh 1974). Naringinase debittering was inhibited by the presence of glucose or fructose, but not by sucrose (Roe and Bruemmer 1977). Enzymatic hydrolysis of naringin to nonbitter naringenin is a two-step process, proceeding first to prunin. Only hydrolysis of prunin to naringenin is inhibited by glucose (Thomas, Smythe, and Laffee 1958); however, since prunin is also bitter, the net effect is an apparent diminution of naringinase activity. By adding β-glucosidase, Roe and Bruemmer (1976) were able to overcome this inhibition.

Depending on fruit size, the albedo of a mature grapefruit can absorb from 100 ml to 150 ml of liquid. Significant quantities of other nutrients can therefore easily be incorporated into the peel. A solution of 7% protein was successfully infused without affecting the naringinase activity, as was a combination of vitamin and mineral additives.

Despite bitterness reduction and flavor enhancement, acceptability of infused peel was marginal, being neither liked nor disliked by the panel (Roe and Bruemmer 1977). Grapefruit albedo, while spongy, is quite resilient and somewhat difficult to reduce by chewing. Some dislike of this product could have occurred as a result of the excessive time required for chewing, which allowed most of the flavoring and sweetener to be extracted. Thus the panelist was left with a residue of unflavored peel that still contained a certain amount of bitterness. This could be mitigated by the use of more persistent sweeteners and flavorings, or by the use of a small amount of pectinase with the naringinase to soften the albedo.

USE OF MICROBIAL ENZYMES TO PEEL AND SECTION CITRUS

Peeling and Production of Sections

Citrus sections comprise the segment juice sacs, usually of grapefruit or orange, excised from segment (carpellary) membranes. Conventional production requires steaming to loosen the peel, machine peeling, lye dipping to remove residual albedo, and rinsing to remove lye and peel residue. Peeled fruit is then sectioned, a labor-intensive process in which a knife is used to cut segments from the carpellary membranes. This step of the process may result in the loss of as much as 40% of the edible fruit as waste juice and rag, and yields sections that, lacking membranes for structural support, are small and delicate. Such sections are of necessity packed in liquid and heat sterilized or cold packed with preservatives.

Vacuum infusion of fungal pectinases has been suggested as a means to both peel and section fruit, avoiding the use of either heat or lye (Bruemmer, Griffin, and Onayemi 1978; Bruemmer 1981). The porous structure of citrus fruit albedo and core is particularly suited to vacuum infusion of degradative enzymes, allowing optimal contact of enzyme solution with substrate. Commercial pectinases are a mixture of pectolytic enzymes, including PG, PME, and PL. These enzymes act to depolymerize pectins of the cell-wall middle lamella, which serve to cement cells together. Degradation of these cementing pectins reduces the adherence between cells, allowing them to be pulled apart easily. In the initial study of this process, sufficient pectinase was infused to cause segment membranes to separate from the juice sacs. Thus sections produced by this new procedure were morphologically similar to cut sections, and shared their fragility. However, enzymatically produced sections were larger than cut sections, and were strongly preferred by a taste panel (Bruemmer, Griffin, and Onayemi 1978). To counteract the fragile nature of such sections, a coating of gelatin was proposed.

Production of Intact Segments

In recent work the enzyme and sectioning process was modified to keep carpellary membranes intact, yielding entire segments (Baker 1987; Baker, Foerster, and Parish 1988; Baker and Bruemmer 1989). Such segments offer three advantages over conventionally produced sections: (1) all of the edible portion of fruit is utilized; (2) segments are larger and more attractive than cut sections; and (3) segments with intact membranes require less severe enzyme treatment and are therefore firm enough to be packed without liquid cover. This latter advantage would allow citrus segments to be sold as a convenient fresh fruit product under the proper storage conditions.

Peeling Efficiency versus Enzyme Activity

Bruemmer, Griffin, and Onayemi (1978) reported that concentrations of commercial pectinases necessary to cause peel to slip easily from fruit after a given holding time varied 30-fold between preparations. The authors saw some correlation between pectin degradative activity of enzyme preparations and peeling efficacy. Baker and Bruemmer (1989) also found that those preparations with the highest PG activity per unit protein were most effective in peeling. However, when concentration levels of preparations were calculated by regression analysis to give equivalent peeling activity, and concentrations of constituent enzymes calculated for those levels, enzyme levels varied widely.

If PG were the sole component contributing to peeling activity, then at equivalent peeling ratings pectinase preparations should contain similar PG activities. They did not, however, with variation in PG activity on polygalacturonic acid (PGA) substrate ranging 160-fold. There was also no correspondence between peeling effectiveness and activities of cellulase, PME, or PG on pectin substrate. Preparations may assay for high activity on pure substrate and under specific conditions of pH and temperature, but be far less effective against a mixed substrate such as citrus peel. A more accurate measure of peeling activity may be possible with an assay based on a substrate of alcohol-insoluble solids of citrus peel, as was suggested by Ben-Shalom and Pinto (1986). This would provide a substrate more closely resembling intact peel than the refined pectin used in standard assays. However, other factors, such as varying pH sensitivities of PG, PL, and PME in different commercial pectinase preparations, complicate the development of an assay that would successfully predict peeling efficacy under all conditions.

Texture of Stored Segments

When infused, sufficient pectinase solution penetrates the carpellary membranes and juice vesicles to effect some softening. Excessive softening of

TABLE 3-4   Segment Firmness, Membrane Toughness, and Liquid Loss
of Stored Segments[a]

Pectinase Water Controls	Vesicle Compression[b] (lb)	Membrane Shear[c] (lb)	Liquid Loss (%)
	239.0	242.0	4.0
1	164.9	151.3	9.4
6	98.7	108.9	13.1
4	107.1	90.9	14.5
7	120.5	88.5	15.9
5	121.9	106.7	16.7
3	96.9	73.0	24.1
2	92.5	82.9	28.0

Source: Data from Baker and Bruemmer 1989.

[a]Segments stored 30 days at 2°C in sealed bags.
[b]Compressed to 3 mm in an Ottowa Texture Measurement System with an Instron Model 1011 Universal Tester.
[c]Measured with a Kramer Shear Cell in an Instron Universal Tester.

vesicles is undesirable, but some reduction of segment membrane toughness, particularly of grapefruit, is advantageous. When pectinases were used at concentrations giving equivalent peeling ratings, firmness of segments (as measured by vesicle compression) and membrane toughness (measured as shear force) varied widely after storage at 2°C for 30 days (Table 3-4). Pectinase infusion reduced firmness of segments from 38% to 69% of water-infused controls, and membrane toughness by 30% to 65% (Baker and Bruemmer 1989). Reductions in segment firmness and membrane toughness were generally comparable for individual pectinases, but there was no correlation between these parameters and specific enzyme activities. When more mature late season fruit were used, pectinase use resulted in excessive softening of segments. A thin coating of an edible alginate film containing calcium as hardener improved firmness of such segments by 89%.

Fluid Loss of Stored Segments

Bruemmer, Griffin, and Onayemi (1978) reported improved texture and no loss of liquid from enzyme-sectioned grapefruit over a three-day period. However, for effective marketing, fresh unpasteurized dry-packed segments should have a refrigerated shelf life of 3 to 4 weeks. When enzyme-peeled and -segmented grapefruit were stored at 2°C for one month, fluid losses ranged from 9.4% to 28%, depending on the pectinase used (Table 3-4) (Baker and Bruemmer 1989). In a similar experiment performed with firmer earlier season fruit, fluid losses varied from 4% to 13%. Excessive loss of liquid from segments caused noticeable flaccidity, and in a dry pack was visually objectionable. Use of early season fruit and selection of the proper pectinase can minimize this defect. Use of an alginate film with calcium salt hardener was relatively ineffective in stemming fluid loss of segments. For example, an 89% increase in segment firmness was accompanied by only a 16% decrease in fluid loss.

Stability of Flavor and Appearance

Fresh, unpasteurized enzyme-peeled segments developed significant off-flavor in 3 weeks if stored in oxygen-permeable polyethylene bags. Flavor of segments stored in moderately permeable bags ($\sim$ 3000 cc O_2/m^2/day) was preferred to flavor of segments stored in either polyethylene or a very effective oxygen barrier bag (4–6 cc O_2/m^2/day. Regardless of the storage atmosphere, a defect affecting both flavor and appearance of stored grapefruit segments was white deposits of naringin crystals on the segment surface (Baker, Foerster, and Parish 1988). Naringin did not appear to be synthesized at the segment surface or transported to the surface from within. Approximately 15–20% of the total segment naringin was localized in residual albedo cells adhering to the carpellary membrane, and visibility was apparently enhanced by selective crystal growth. Vigorous water misting, soft scrubbing, or brief immersion in dilute NaOH were all effective in reducing subsequent growth of naringin crystals on stored segments.

Microbial Stability

If manufactured and distributed as a fresh unpasteurized product, segments would be susceptible to microbial contamination. In a study of grapefruit segments stored at 2°C for 9 weeks in polyethylene bags, Baker and Bruemmer (1989) found that bacterial contamination was minimal, and declined during storage. Initial yeast contamination was also minimal, but accelerated rapidly after 3 weeks of storage. A dip in 0.2% potassium sorbate completely inhibited this growth. Further studies (Baker and Parish, unpublished) have shown that packaging in bags of low oxygen permeability is as effective as sorbate in suppressing microbial growth. Thus several potential means of controlling microbial contamination in fresh segments appear feasible. For short-term storage (less than 4 weeks) of acidic segments, such as grapefruit, maintenance at 2°C may be sufficient to suppress contamination. For longer storage periods or less acidic fruit, dipping in a preservative or packaging in an oxygen-barrier film would inhibit spoilage. Since flavor stability is also dependent on reduced oxygen packaging, the latter procedure would accomplish both purposes.

Production of Whole Peeled Fruit

By proper adjustment of enzyme activity, grapefruit and oranges can be effectively peeled without damaging the segment membranes. Sufficient adherence exists between segments to permit an alternative marketing form, that of whole peeled fruit (Baker 1987). Intact peeled fruit offer the consumer the option of point-of-use generation of segments, or particularly with oranges, the production of slices for garnish. Whole peeled grapefruit suffered less fluid loss during storage (0–5.7%) than did fruit separated into segments (3–13%). Shrink-filming individual fruit with an oxygen-barrier film would provide flavor and microbial stability, and protection against disaggregation of segments. Whole peeled citrus are currently being marketed in institutional packages (Bush 1990).

BIOCONTROL OF POSTHARVEST DISEASES

Problems with Chemical Control of Postharvest Diseases

Conservative estimates of postharvest losses of fruits and vegetables from spoilage are at around 24% in the United States and 50% worldwide. Often losses occur due to rots caused by microorganisms (Wilson and Wisniewski 1989). For example, *Penicillium digitatum* (green mold) and *P. italicum* (blue mold) are responsible for most decay losses of citrus worldwide (Bancroft et al. 1984). Currently, fungicides are the primary means of controlling postharvest diseases, but use of pesticides is becoming unpopular with consumers (Wilson and Chalutz 1989). The major fungicides used to control postharvest diseases of fruits and vegetables are under review by the U.S. Environmental Protection Agency, and already there is an increase in legislation against the use of some of these chemicals (Jeffries and Jeger 1990). This illustrates the need to find alternative methods of disease control.

Alternatives to Chemical Control

Fruits and vegetables have antimicrobial constitutive and inducible compounds that are potential biocontrol agents, including plant volatiles and essential oils as well as naturally occurring antagonistic microorganisms that exist on fruit and vegetable surfaces (Wilson and Wisniewski 1989). Acetaldehyde vapor has been shown to control postharvest decay of apple (Stadelbacher and Prasad 1974), while benzaldehyde and other stone fruit volatiles inhibited growth of fungi responsible for fruit rots (Wilson, Franklin, and Otto 1987). Essential oils from leaves of several species of plants have been reported to boost resistance of several stored commodities to attack by *Aspergillus flavus* and *A. versicolor* (Wilson and Wisniewski 1989). Antagonistic microorganisms, however, have attracted the most attention and shown the most promise. Most of these microorganisms apparently produce antibiotics that inhibit the growth of target pathogens. This may be the reason that washed fruits and vegetables are most susceptible to decay than those that are unwashed (Wilson and Wisniewski 1989). These antagonistic microorganisms, including yeast, fungi, and bacteria, apparently inhibit growth of pathogenic organisms by producing antibiotic compounds such as iturin and pyrrolnitrin (Wilson 1989). Alternatively, it is thought that some of these antagonists hold growth of pathogens in check by aggressively competing for nutrients and space (Droby et al. 1989; Wilson 1989).

Examples of Biological Control

Recently, biological control of *Botrytis* has been demonstrated on strawberries (Tronsmo 1986) and grapes (Dubos 1987) using various *Trichoderma* species, and on apples using strains of the yeast, *Candida* sp. (McLaughlin et al. 1990) and *Cryptococcus laurentii* (Roberts 1990). In the case of *Candida* sp., control was enhanced by applying the yeast in aqueous salt solutions compared to application of yeast in aqueous solutions without salt or salt solutions alone. How the salt solutions enhanced control with yeasts has not yet been determined. Control of brown rot (*Monilinia fructigena*) of peaches and other stone fruit was inhibited by use of a bacterium *Bacillus subtilis* (Wilson and Pusey 1985; Pusey et al. 1986). Control of *Botrytis cinerea* using both yeasts and bacteria was reported for apples (Janisiewicz 1987) and pears (Janisiewicz and Roitman 1988). Green and blue mold infection (*Penicillium digitatum* and *italicum,* respectively) of citrus was effectively inhibited by antagonistic yeasts (*Debaryomyces hansenii* and *Aureobasidium pullulans*) and bacteria (*Pseudomonas cepacia* and *syringae*) with only *P. cepacia* producing antibiotoc zones against the Penicillia rot organisms in culture (Wilson and Chalutz 1989). The other antagonists may have inhibited the growth of the Penicillia rot organisms by competing for nutrients (Droby et al. 1989). Competition for nutrients or site exclusion may also be the case for the control of *Botrytis cinerea* in apple by the yeast *Cryptococcus laurentii* (Roberts 1990) and *Candida* sp. (McLaughlin et al. 1990) as well as anthracnose of mangoes by a postharvest dip in a *Pseudomonas* suspension (Koomen et al. 1990). This type of control, as opposed to the production of antibiotics, could be advantageous because antibiotic-producing agents could be classified as agrochemicals, in terms of toxicological and environmental testing, which would delay clearance for commercial use (Jeffries and Jeger 1990).

Future Research in the Area of Biocontrol of Postharvest Diseases

Assessment of Current Pre- and Postharvest Practices

Management of epiphytic microorganisms on fruit and vegetable surfaces could lead to enhanced resistance to postharvest diseases. The effect of preharvest practices, such as the use of nutrient and pesticide sprays, on fruit surface microflora needs to be investigated and perhaps modified (Wilson 1989). For example, preharvest sprays of nutrient-rich materials have been reported to stimulate the development of indigenous microorganisms on fruit and leaf surfaces, whereas preharvest fungicidal sprays reduced their population (Jeffries and Jeger 1990). Similar studies need to be conducted on the postharvest use of Clorox washes, hot dips, and waxes for their effect on the epiphytic microorganism populations living on fresh produce (Wilson 1989). The goal of such research would be to enhance the growth of antagonist species on fruit and vegetable surfaces in order to increase disease resistance during postharvest storage.

Direct Application of Antagonists to Fruits

Antagonist populations can also be boosted by the addition of antagonists from an external source (Jeffries and Jeger 1990). This "bioaugmentation" approach (Jeffries and Jeger 1990) can be used at the preharvest stage, as has been demonstrated by using preharvest sprays of *Trichoderma* to control *Botrytis* disease of strawberries (Tronsmo 1986) and grapes (Dubos 1987), or postharvest, as has been reported using postharvest dips of *Bacillus subtilis* to control brown rot of peaches (Pusey et al. 1988). Some natural antagonists may be under genetic control of the host, and as such could be managed through plant breeding and selection (Wilson 1989).

Commercial Adaptation of Biocontrol Methods

Successful application of these antagonists on a commercial scale is under investigation. Some success has been achieved by incorporation of the antagonist *Bacillus subtilis* into a wax normally used to coat peaches (Pusey et al. 1988). Brown rot was effectively controlled in this manner, but considerable variation in disease control was encountered with different preparations of the antagonist (Wilson and Wisniewski 1989). More research is needed on methods of antagonist application through additives such as nutrients, wetting agents (Wilson and Wisniewski 1989), preharvest sprays, postharvest dips (Jeffries and Jegger 1990), dump tanks (Wilson 1989), or edible coatings. To date, these agents have not been adopted commercially. The fresh produce industry is resistant to change while effective fungicides are still available and there is still a market for chemically treated produce.

References

Baker, R. A. 1987. Preparation and storage stability of enzyme peeled grapefruit. Paper read at the 1987 Subtropical Technology Conference, Winter Haven, FL.

Baker, R. A., and J. H. Bruemmer. 1989. Quality and stability of enzymically peeled and sectioned citrus fruit. In *Quality Factors of Fruits and Vegetables—Chemistry and Technology,* ACS Symposium Series 405, ed. J. J. Jen, pp. 140–148. Washington, DC: American Chemical Society.

Baker, R. A., J. A. Foerster, and M. E. Parish. 1988. Flavor, texture, and microbial stability of enzyme peeled citrus. Paper read at the 1988 Subtropical Technology Conference, Winter Haven, FL.

Baldwin, E. A., and R. H. Biggs. 1988. Cell-wall lysing enzymes and products of cell-wall digestion elicit ethylene in citrus. *Physiol. Plant.* 73:58–64.

Baldwin, E. A., and R. Pressey. 1988a. Tomato polygalacturonase elicits ethylene production in tomato fruit. *J. Am. Soc. Hort. Sci.* 113(1):92–95.

Baldwin, E. A., and R. Pressey. 1988b. Treatment of tomatoes with an exo-enzyme increases ethylene and accelerates ripening. *Proc. Fla. State Hort. Soc.* 101:215–217.

Baldwin, E. A., and R. Pressey. 1989. Pectic enzymes in Pectolyase: Separation, characterization, and induction of ethylene in fruits. *Plant Physiol.* 90:191–196.

Baldwin, E. A., and R. Pressey. 1990. Exopolygalacturonase elicits ethylene production in tomato. *HortScience* 25(7):779–780.

Baldwin, E. A., M. O. Nisperos-Carriedo, and M. G. Moshonas. 1991. Quantitative analysis of flavor and other volatiles and for certain constituents of two tomato cultivars during ripening. *J. Am. Soc. Hort. Sci.* 116:265–269.

Bancroft, M. N., P. D. Gardner, J. W. Eckert, and J. L. Baritelle. 1984. Comparison of decay control strategies in California lemon packinghouses. *Plant Dis.* 68:24–28.

Ben-Shalom, N., and R. Pinto. 1986. Pectolytic enzyme studies for peeling of grapefruit segment membrane. *J. Food Sci.* 51:421–423.

Biale, J. B., and R. E. Young. 1981. Respiration and ripening-fruits retrospect and prospects. In *Recent Advances in the Biochemistry of Fruit and Vegetables,* ed. J. Friend and M. J. C. Rhodes, pp. 1–39. London: Academic Pres.

Brecht, J. H., and D. J. Huber. 1988. Products released from enzymically active cell wall stimulate ethylene production and ripening in preclimacteric tomato (*Lycopersicon esculentum* Mill.) fruit. *Plant Physiol.* 88:1037–1041.

Bruemmer, J. H. 1981. Method of Preparing Citrus Fruit Sections with Fresh Fruit Flavor and Appearance. U.S. Patent No. 4,284,651.

Bruemmer, J. H., A. W. Griffin, and O. Onayemi. 1978. Sectionizing grapefruit by enzyme digestion. *Proc. Fla. State Hort. Soc.* 91:112–114.

Bush, P. 1990. When less is more. *Prepared Foods* 159(9):138–140.

DeLaat, A. M., and L. C. VanLoon. 1982. Regulation of ethylene biosynthesis in virus-infected tobacco leaves. *Plant Physiol.* **69**:240–245.

Droby, S., E. Chalutz, C. L. Wilson, and M. Wisniewski. 1989. Characterization of the biocontrol activity of *Debaryomyces hansenii* in the control of *Penicillium digitatum* on grapefruit. *Can. J. Microbiol.* **35**(8):794–800.

Dubos, B. 1987. Fungal antagonism in aerial agrobiocenoses. In *Innovative Approaches to Plant Disease Control,* ed. I. Chet, pp. 107–135. New York: Wiley.

Evenson, K. B., M. G. Bausher, and R. H. Biggs. 1981. Wound-induced ethylene production in peel explants of 'Valencia' orange fruit. *HortScience* **16**:43.

Grierson, D. 1985. Gene expression in ripening tomato fruit. *CRC Critical Rev. Plant Sci.* **3**(2):113–132.

Grierson, D., and G. A. Tucker. 1983. Timing of ethylene and polygalacturonase synthesis in relation to the control of tomato fruit ripening. *Planta* **157**:174.

Guadagni, D. G., V. P. Maier, and J. H. Turnbaugh. 1974. Some factors affecting sensory thresholds and relative bitterness of limonin and naringin. *J. Sci. Food Agric.* **25**:1199–1205.

Ishii, S. 1976. Enzymatic maceration of plant tissues by endo-pectin lyase and endo-polygalacturonase from *Asperigillus japonicus*. *Phytopathology* **66**:281–289.

Ishii, S., and T. Yokotsuka. 1975. Purification and properties of pectin lyase from *Aspergillus japonicus*. *Agric. Biol. Chem.* **39**:313–321.

Janisiewicz, W. J. 1987. Postharvest biological control of blue mold on apples. *Phytopathology* **77**:481–485.

Janisiewicz, W. J., and J. Roitman. 1988. Biological control of blue mold and gray mold on apple and pear with *Pseudomonas cepacia*. *Phytopathology* **78**:1697–1700.

Javeri, H., R. Toledo, and L. Wicker. 1990. Effect of vacuum infusion of citrus pectinmethylesterase and calcium on firmness of peaches. *J. Food Sci.* **56**:739–742.

Jeffery, D., C. Smith, P. Goodenough, I. Prosser, and D. Grierson, 1984. Ethylene-independent and ethylene-dependent biochemical changes in ripening tomatoes. *Plant Physiol.* **74**:32–38.

Jeffries, P., and M. J. Jeger. 1990. The biological control of postharvest diseases of fruit. *Postharvest News and Info.* **1**:365–368.

Ketring, D. L., and H. A. Melouk. 1982. Ethylene production and leaflet abscission of three peanut genotypes infected with *Cercospora arachidicola* Hori. *Plant Physiol.* **69**:789–792.

Khudairi, A. K. 1972. The ripening of tomatoes. *Amer. Scientist* **60**:696–707.

Kim, J., K. C. Gross, and T. Solomos. 1987. Characterization of the stimulation of ethylene production by galactose in tomato (*Lycopersicon esculentum* Mill.) fruit. *Plant Physiol.* **85**:804–807.

Koomen, I., J. C. Dodd, M. J. Jeger, and P. Jeffries. 1990. Postharvest biocontrol of anthracnose disease of mangoes. *J. Sci. Food Agric.* **50**:137–138.

McFeeters, R. F. 1989. Function of metal cations in regulating the texture of acidified vegetables. In *Quality Factors of Fruits and Vegetables—Chemistry and Technology,* ACS Symposium Series 405, ed. J. J. Jen, pp. 125–139. Washington, DC: American Chemical Society.

McLaughlin, R. J., M. E. Wisniewski, C. L. Wilson, and E. Chalutz. 1990. Effect of inoculum concentration and salt solutions on biological control of postharvest diseases of apple with *Candida* sp. *Phytopathology* **80**:456–461.

McNeil, N., A. G. Darvill, S. C. Fry, and P. Albersheim. 1984. Structure and function of the primary cell walls in plants. *Ann. Rev. Biochem.* **53**:625–663.

Mizrahi, Y., H. Dostal, and J. Cherry. 1975. Ethylene-induced ripening in attached '*rin*' fruits, a non-ripening mutant of tomato. *HortScience* **10**:414–415.

Pressey R. 1986. Polygalacturonases in higher plants. In *Chemistry and Function of Pectins*, ACS Symposium Series 310, ed. J. Jen, pp. 157–174. Washington, DC: American Chemical Society.

Pressey, R. 1987. Exopolygalacturonase in tomato fruit. *Phytochem.* **26**:1867–1870.

Pressey, R., and J. K. Avants 1977. Occurrence and properties of polygalacturonase in *Avena* and other plants. *Plant Physiol.* **60**:548–553.

Pusey, P. L., C. L. Wilson, M. W. Hotchkiss, and J. D. Franklin. 1986. Compatibility of *Bacillus subtilis* for postharvest control of peach brown rot with commercial fruit waxes, dicloran and cold-storage conditions. *Plant Dis.* **70**:587–590.

Pusey, P. L., M. W. Hotchkiss, H. T. Dulmage, R. A. Baumgardner, E. I. Zehr, et al. 1988. Pilot test for commercial production and application of *Bacillus subtilis* (B-3) for postharvest control of peach brown rot. *Plant Dis.* **72**:622–626.

Rexova-Benkova, L., and O. Markovic. 1976. Pectic enzymes. *Adv. Carbohydr. Chem. Biochem.* **33**:323–385.

Roberts, R. G. 1990. Postharvest biological control of gray mold of apple by *Cryptococcus laurentii. Phytopathology* **80**:526–530.

Roe, B., and J. H. Bruemmer. 1976. New grapefruit product: Debitterizing albedo *in situ. Proc. Fla. State Hort. Soc.* **89**:191–194.

Roe, B., and J. H. Bruemmer. 1977. Treatment requirements for debittering and fortifying grapefruit and stable storage of the product. *Proc. Fla. State Hort. Soc.* **90**:180–182.

Sinclair, W. B. 1972. *The Grapefruit, Its Composition, Physiology, and Products.* University of California, Division of Agricultural Science. University of California Press, Riverside, CA, p. 137.

Stadelbacher, G. J., and K. Prasad. 1974. Postharvest decay control of apple by acetaldehyde vapor. *J. Am. Soc. Hort. Sci.* **99**:364–368.

Thomas, D. W., C. V. Smythe, and M. D. Laffee. 1958. Enzymatic hydrolysis of naringin, the bitter principle of grapefruit. *Food Res.* **23**:591–598.

Tong, C., J. Labavitch, and S. F. Yang. 1986. The induction of ethylene production from pear cell culture by cell wall fragments. *Plant Physiol.* **81**:929–930.

Tronsmo, A. 1986. Use of *Trichoderma* spp. in biological control of necrotrophic pathogens. In *Microbiology of the Phyllosphere.* ed. N. J. Fokkema and J. van den Heuvel, pp. 348–362. Cambridge, UK: Cambridge University Press.

Wills, R. H., T. H. Lee, D. Graham, W. B. McGlasson, and E. G. Hall. 1981. *Postharvest: An Introduction to the Physiology and Handling of Fruit and Vegetables.* Westport, CT: AVI.

Wilson, C. L. 1989. Managing the microflora of harvested fruits and vegetables to enhance resistance. *Phytopathology* **79**:1387–1390.

Wilson, C. L., and E. Chalutz. 1989. Postharvest biological control of *Penicillium* rots of citrus with antagonistic yeasts and bacteria. *Scientia Hort.* **40:**105–112.

Wilson, C. L., and P. L. Pusey. 1985. Potential for biological control of postharvest plant diseases. *Plant Dis.* **69:**375–378.

Wilson, C. L., and M. E. Wisniewski. 1989. Biological control of postharvest diseases of fruits and vegetables: An emerging technology. *Ann. Rev. Phytopathol.* **27:**425–441.

Wilson, C. L., J. D. Franklin, and B. Otto. 1987. Fruit volatiles inhibitory to *Monilinia fructicola* and *Botrytis cinerea*. *Plant Dis.* **71:**316–319.

4

Safety Evaluation of Food Enzymes from Genetically Engineered Organisms

Nancy W. Zeman and W. Martin Teague

Over the years, the evaluation of food safety has changed from simple trial and error sampling to a system of scientific screening and testing procedures that have accumulated from knowledge and experience gained in such areas as plant and animal breeding and the use of microorganisms in food products. During the development of a new food or food ingredient, the developer is responsible for evaluating and documenting the safety of the product. The safety evaluation must satisfy all applicable legal and regulatory requirements in the country of the substance's intended use. Because cultures and public trust in scientific procedures vary from nation to nation, so will the legal and regulatory requirements.

The recent use of recombinant DNA (rDNA) technology in the production of food and food ingredients has generated a challenge for those responsible for food safety evaluation and regulatory approval processes. Because rDNA techniques allow the transfer of genetic information between unrelated species or even different biological kingdoms, and because these techniques involve laboratory manipulations that occur outside of the living organisms, there was concern that regulatory agencies would not be able to regulate rDNA-derived foods under existing law. Such uncertainty increased the business risk of investing in developing products where these techniques were used.

This chapter reviews several examples of how the current U.S. federal regulatory requirements for introduction of a new food into commerce have been applied to enzymes produced by genetically engineered organisms. A brief description of current, pertinent regulations is given, safety evaluation aspects of these enzymes are discussed, and expectations for the future are outlined. Other nations have addressed the regulation of rDNA products in the food industry; however, a discussion of these issues is beyond the scope of this chapter.

BACKGROUND

In the United States, most food and food ingredients are regulated by the Food and Drug Administration (FDA) under the federal Food, Drug, and Cosmetic Act (1982). Food substances can be placed into one of the following three classifications:

1. Substances that are generally recognized as safe, or GRAS, for their intended use.
2. FDA-approved food additives, or substances that have been shown to be safe for specific uses and at specific levels, but are not generally recognized as safe.
3. Unapproved food additives, considered added poisonous or deleterious substances that may render a food substance injurious to health.

A GRAS ingredient is defined as a substance that is generally recognized among experts as having been adequately shown to be safe under the conditions of its intended use. Eligibility for GRAS status may be based either on a safe history of common use in food prior to January 1, 1958, or on scientific information and procedures. General recognition of safety based upon scientific procedures ordinarily requires published studies, which may be corroborated by unpublished studies and other data and information, which indicate that the scientific community considers the substance safe for food use.

A new food ingredient may be designated as GRAS in one of two ways. First, the manufacturer may make a self-determination of GRAS status by establishing to its own satisfaction that the requisites for general recognition of safety have been met. The substance may be used without seeking FDA affirmation; however, the manufacturer runs the risk that the FDA may reach a different conclusion. The agency may challenge use of the new substance on the grounds that it is not GRAS and thus is an unapproved food additive. A food containing an unapproved food additive is considered adulterated and, thus, unlawful. Because of these uncertainties and regulatory risks, most companies are reluctant to "self-certify" new products as GRAS. A second way to achieve GRAS status is for a manufacturer to seek FDA approval for use of a new food ingredient through the filing of a GRAS affirmation petition. Requirements for obtaining FDA affirmation of GRAS status are outlined in Title 21, Code of Federal Regulations 170.35. Although the actual FDA affirmation and publication of regulations for manufacture and use of the new food substance may take years to obtain, it is not uncommon for enzyme manufacturers to begin marketing of a new product after the affirmation petition has been accepted for filing by the agency, a notice of filing has appeared in the *Federal Register,* and the period for public comment has passed without adverse comment.

The enzymes listed in Table 4-1 are either affirmed as GRAS or are the subject of GRAS affirmation petitions that are in the process of being reviewed. In general, GRAS status (versus food additive status) is preferred by the enzyme industry since, if a substance is GRAS, the regulations covering its use will generally be broader and the substance can be used without notifying the agency of its use.

The application of genetic engineering to production of food ingredients

TABLE 4-1   Enzymes Generally Recognized as Safe by the FDA or the Subject of GRAS Petitions

Common Name	Source
α-Amylase	*Aspergillus niger, Aspergillus oryzae, Bacillus licheniformis, Bacillus stearothermophilus, Bacillus subtilis,* barley malt, or *Rhizopus oryzae*
Aminopeptidase	*Lactobacillus lactis*
β-Amylase	Barley malt
Bromelin	*Ananas comosus* or *Ananas bracteatas* L.
Catalase	*Aspergillus niger* or bovine liver
Cellulase	*Aspergillus niger* or *Trichoderma longibrachiatum*
Ficin	*Ficus* spp.
Glucoamylase	*Aspergillus niger* or *Aspergillus oryzae*
β-Glucanase	*Aspergillus niger* or *Bacillus subtilis*
Glucose isomerase	*Actinoplanes missouriensis, Bacillus coagulans, Microbacterium arborescens, Streptomyces murinus, Streptomyces olivaceus, Streptomyces olivochromogenes,* or *Streptomyces rubiginosus*
Glucose oxidase	*Aspergillus niger*
Hemicellulase	*Aspergillus niger*
Invertase	*Saccharomyces* spp.
Lactase	*Aspergillus niger, Aspergillus oryzae, Candida pseudotropicalis,* or *Kluyveromyces* spp.
Lipase	*Aspergillus niger; Aspergillus oryzae;* forestomach of calves, kids, or lambs; or bovine or porcine pancreatic tissue
Lipase-esterase	*Mucor miehei*
Maltogenic amylase	*Bacillus subtilis*
Papain	*Carica papaya* L.
Pectinase	*Aspergillus niger*
Pepsin	Bovine or porcine stomachs
Phospholipase A2	Pancreatic tissue
Protease	*Aspergillus niger, Aspergillus oryzae,* or *Bacillus subtilis*
Pullulanase	*Bacillus acidopullulyticus*
Rennet (chymosin)	*Aspergillus niger* var. *awamori, Escherichia coli* K-12, *Kluyveromyces marxianus* var. *lactis,* or fourth stomach of ruminants
Trypsin	Pancreatic tissue

TABLE 4-2   GRAS Petitions Accepted for Filing by the FDA for rDNA-derived Products

Common Name	Source	Host	Company	FDA Acceptance Date
α-amylase	*Bacillus stearother- mophilus*	Bacillus sub- tilis	CPC Interna- tional, Inc.	March 27, 1986
Chymosin	Synthetic DNA	*Escherichia coli*	Pfizer, Inc.	February 9, 1988
α-amylase	*Bacillus meg- aterium*	*Bacillus sub- tilis*	Enzyme Bio- Systems Ltd.	May 5, 1988
Chymosin	Calf stomach	*Kluyveromyces marxianus* var. *lactis*	Gist- Brocades, Inc.	May 10, 1989
Chymosin	Calf stomach	*Aspergillus ni- ger* var. *awa- mori*	Genencor, Inc.	September 22, 1989
Maltogenic amylase	*Bacillus stearother- mophilus*	*Bacillus sub- tilis*	Novo Labora- tories, Inc.	January 12, 1990

raised the question: could a food substance produced by rDNA techniques be considered GRAS (McNamara 1987)? The FDA's response was that any product satisfying GRAS criteria could be eligible for GRAS status. A rDNA-derived product may be considered as GRAS on the basis of scientific evidence; the method of manufacture did not preclude it from GRAS consideration (Miller and Thompson 1985). The agency verified this statement when it accepted for filing on March 27, 1986, the first GRAS petition for a rDNA-derived food ingredient (CPC International 1986), an α-amylase from *Bacillus stearothermophilus* produced by a recombinant *Bacillus subtilis*. Since then, an additional five GRAS petitions where rDNA technology was used, all for enzymes, have been accepted for filing (see Table 4-2). One of these enzymes, a chymosin derived from *Escherichia coli,* has been affirmed as GRAS (Food and Drug Administration 1990a) by the FDA.

GRAS PETITIONS FOR rDNA-DERIVED ENZYMES

It is not surprising that rDNA-derived enzymes would be the first category of food ingredients for which GRAS petitions would be filed. Enzymes have had a long history of safe use in the food industry, and enzyme manufacturers are experienced in collecting safety information and securing GRAS status for these substances. Furthermore, enzymes are relatively simple food ingredients. Most commercial enzymes, especially those from microbial sources, are fairly small, extracellular proteins that frequently are not glycolysated. This makes them ideal candidates for genetic manipulation.

Much of the safety information that has been provided in GRAS petitions for enzyme products derived from genetically engineered organisms is no different from that provided in GRAS petitions for enzymes derived from "natural" organisms. All petitions for enzymes of microbial origin include a literature review to show the safety of the enzyme and the host organism, feeding studies in animal models with the enzyme product, and data indicating conformation of product properties with Food Chemicals Codex (FCC) specifications (National Academy of Sciences 1981).

General FCC specifications require that the enzyme preparation should be produced in accordance with good manufacturing practices (Title 21, Code of Federal Regulations, Part 110), the product should cause no increase in the total microbial count in the treated food over the level accepted for the respective food, the culture conditions used to produce the enzyme should be such that a controlled fermentation is ensured, and the product should not contain mycotoxins. Specific FCC requirements address levels of arsenic, heavy metals, coliforms, and *Salmonella* sp. in the product.

The areas in petitions where expanded safety information for rDNA-derived enzymes (over non-rDNA-derived enzymes) has been provided are in the description of the organism, the product identity, and in the environmental assessment. The description of the genetically engineered organism usually includes information on the host, sources of any donor DNA, sources of all plasmids used, and the DNA sequence for any inserted DNA. Attempts to eliminate unnecessary DNA, particularly known coding regions, are generally made to help assure that no undesired proteins are made. The presence of antibiotic resistance genes in final constructs is a topic that has received a lot of attention. Antibiotic resistance markers are frequently contained on the commonly used plasmids and are essential for screening during the cloning processes. The Agency's major concern is that DNA that encodes for these genes, if present in final product, could be passed on to human microflora after ingestion, or that the genes for the antibiotic resistances could be released and exchanged with other microorganisms in the environment. In the petitions filed to date, three approaches have been used to overcome these concerns. In certain instances, the resistance markers have been removed or are inactive in the final host construction. For those cases where the antibiotic resistance is maintained, either the DNA that encodes antibiotic resistance is destroyed during postfermentation enzyme processing, so that no transformable DNA is present in the product, or the antibiotic resistance marker does not code for an antibiotic that is commonly used in human therapeutics.

In the cases where petitions have been filed for enzymes that are already GRAS (or the subject of GRAS petitions), but are to be produced by a genetically engineered microorganism (e.g., chymosin), data are given to show that the rDNA-derived enzyme is "identical" to the enzyme from its natural host. Physical, chemical, and functional properties have all been used to demonstrate "nature-identical" products.

Inclusion of an environmental assessment into GRAS petitions is a recent requirement established by the National Environmental Policy Act (Food and Drug Administration 1985). For products from rDNA organisms, the environmental assessments have shown minimized release of the genetically engineered host organism by containment within the production facility. Details of the physical and biological containment procedures are provided. A discussion of the fate of the organisms in the environment, including a description of the chance for genetic exchange to occur and consequences, is given. A demonstration that no production host cells and no transformable DNA are contained in the final product is generally required.

It is interesting to note that of the six GRAS petitions that have been filed for enzymes produced by the new biotechnology, three are for amylases and three are for chymosin. The fact that the enzyme industry has, thus far, concentrated on these two types of enzymes with a large share of the food enzyme market, is a comment on the perceived high business risk that was associated with these projects when they were initiated. The submission of four of the first six petitions has answered an early question that was asked concerning the FDA's requirements for regulating rDNA-derived food products. Would a substance produced by recombinant technology that is "identical" to a substance already affirmed as GRAS also be GRAS (Gibbs and Kahan 1986), or would a new petition be required? The acceptance for filing of these submissions is consistent with the FDA's indication that new petitions were necessary even when a product derived from rDNA organisms was identical to a previously approved substance produced in a conventional way (Miller and Thompson 1985). However, there are indications that the agency might relax its position on these types of products in the future (Berkowitz and Maryanski 1989). Details of the individual petitions submitted to date are summarized below.

Bacillus stearothermophilus α-amylase Derived from *Bacillus subtilis*

As mentioned previously, the petition submitted by CPC International, Inc., for affirmation of GRAS status for a *Bacillus stearothermophilus* α-amylase produced in *Bacillus subtilis,* was the first petition for a rDNA-derived food ingredient to be accepted for filing by the FDA (CPC International 1986). Although a GRAS petition for the α-amylase produced by the natural *B. stearothermophilus* host had been previously accepted for filing in 1983 (CPC International 1983), a petition affirming the safety of the same enzyme derived from a genetically engineered host was required. General information concerning the safety issues was resubmitted with the new petition. For the identity of the ingredient, information was presented demonstrating that the cloned α-amylase was equivalent to that from *B. stearothermophilus* (Zeman and McCrea 1985). The functionality in foods and dietary exposure remained the same since no new markets were anticipated for this product. The microbiological review included the details of the cloning of the α-amylase gene. The gene, plasmid borne in *B. stearothermophilus,* was first transferred into an *Escherichia coli* K-12 strain using the vector pBR327. The location of the gene within the piece of inserted DNA was identified and unnecessary *B. stearothermophilus* DNA was deleted. The plasmid pUB110 was used to clone the gene into an asporogenic strain of *B. subtilis.* Approximately 70% of the original pUB110 plasmid DNA was removed, including genetic information that encoded resistance to the antibiotics kanamycin and bleomycin. The final plasmid contained approximately 40% *B. stearothermophilus* DNA that coded for the structural gene plus the necessary information for expression and secretion. The remaining 60% of the plasmid was derived from the *B. subtilis* cloning vector pUB110 (Zeman and McCrea 1985). All rDNA-associated laboratory work was done following the National Institutes of Health guidelines for containment (National Institutes of Health 1986).

Information provided to the agency included a taxonomic identification of the host and donor strains. Evidence was provided showing that the *E. coli* strain used as an intermediate host was nonpathogenic and nontoxicogenic. Assays by reaction to a susceptible cell line (Vero cells) (Konowalchuk, Speirs, and Stavric 1977) were performed to indicate the absence of Shiga-like toxin production. Sequencing and restriction enzyme mapping data were used to show that no *E. coli* pBR327 DNA was present in the final strain and that the pBR327 DNA present in an intermediate plasmid was not sufficient to code for a gene.

Sequence data were used to define the function of the *B. stearothermophilus* DNA present in the final production plasmid. Although the pUB110 plasmid is a commonly used *B. subtilis* cloning vector, it originated from a clinical *Staphylococcus aureus* isolate. Via restriction mapping and sequence data combined with comparisons of the data to literature reports, it was shown that the remaining pUB110 plasmid DNA would pose no human health hazard. Additionally, *S. aureus* enterotoxin tests conducted on the final product were negative.

Evidence was presented showing that the pure, asporogenic production culture was genetically stable during the fermentative production of the α-amylase. Tests were performed to demonstrate the lack of antibiotic production and pathogenicity of the final construct. Standardization of the product was presented to show that it met or exceeded the FCC requirements (National Academy of Sciences 1981).

Five animal feeding studies were performed in accordance with FDA guidelines for safety assessment of food additives and with the FDA's Good Laboratory Practices regulations for nonclinical laboratory studies (Food and Drug Administration 1982). These studies included an acute oral toxicity test in rats, palatability studies in dogs and rats, a subchronic oral toxicity study in dogs, and a subchronic oral toxicity study in in utero-exposed F_1 rats. No untoward effects were observed. These studies were essentially identical to those submitted for the α-amylase derived from the natural host (MacKenzie et al. 1989b). The use of rDNA techniques did not significantly affect the level of toxicological information required for the safety evaluation.

Bacillus megaterium Amylase Derived from Bacillus subtilis

The petition for *Bacillus megaterium* amylase derived from *Bacillus subtilis,* filed by Enzyme Bio-Systems Ltd. (1988), a subsidiary of CPC International, is structured similarly to the cloned *B. stearothermophilus* α-amylase petition. An amylase of unique utility was moved from the *B. megaterium* donor into an asporogenic *B. subtilis* host (Metz et al. 1988) using similar cloning techniques as outlined in the previous petition (see the preceding discussion). Except for the expanded microbiological review and environment assessment, the information regarding product safety accepted for filing in 1988 was not significantly different from that of a food product derived from a nonrecombinant organism. The type of toxicological information submitted was similar to that given in the previous petition (MacKenzie et al. 1989a).

Chymosin Derived from Escherichia coli K-12

Chymosin, a rennin, is used worldwide by the cheese industry to produce a variety of cheeses. This enzyme as derived from calf stomach has a long history of safe use in the food industry and has been affirmed as GRAS by the FDA. Genetic engineering provides a method for producing this enzyme less expensively by using microorganisms instead of animal material as the source. Currently, three petitions for a microbially produced chymosin have been accepted for filing.

In the GRAS petition for chymosin submitted by Pfizer, Inc. (1988), the genetic information for a nature-identical prochymosin was chemically synthesized and inserted in a modified pBR322 plasmid vector containing an ampicillin resistance marker. The plasmid was transferred into a non-pathogenic and nontoxicogenic strain of *E. coli* K-12. After growth of the organism, the prochymosin is converted via autocatalytic cleavage at reduced pH to active chymosin that is identical to calf stomach chymosin as characterized by molecular structure analyses and chemical, functional, and physical properties. The type and amount of safety information contained in the Pfizer petition were similar to that provided in the two GRAS petitions outlined previously. Within the microbiological review, in addition to documenting fully the history, construction, and control of the organism, a Shiga-like toxin test plus a high-dose (5% of diet), five-day feeding study in dogs were performed to establish a 30,000-fold safety margin. Based on the safety information submitted, including evidence that neither the organism nor the plasmid was present in the product, a one-month oral gavage study in rats was submitted as toxicological information. The study showed that the product had no significant effects on food consumption, body weight gain, or clinical pathology parameters in rats.

The chymosin enzyme preparation derived from *E. coli* K-12 was affirmed as GRAS by the FDA as a direct food substance on March 23, 1990 (Food and Drug Administration 1990*a*). This is the first food product from a genetically engineered organism to be approved by the FDA for human consumption.

Chymosin Derived from *Kluyveromyces marxianus* var. *lactis*

Gist-Brocades, Inc. (1989) submitted the second GRAS petition for rDNA-derived chymosin; the genetic information for prochymosin was cloned from the stomach cells of a preruminant calf into *Kluyveromyces marxianus* var. *lactis* using *E. coli* as an intermediate host (van den Berg et al. 1990). The messenger RNA was isolated from the calf cells and transferred first into *E. coli*. The final plasmid in *K. lactis* contained a portion of the *E. coli* vector pUC18, the calf prochymosin coding sequence, a host-specific lactase promoter and terminator, a secretion signal from *Saccharomyces cerevisiae*, and an aminoglycoside resistance marker. Information contained in the submission provided evidence that a pure culture fermentation of the genetically improved microorganism produced a nature-identical chymosin (Meisel 1987, 1988).

Eight toxicological examinations were submitted testing both the enzyme product and cheese made using the enzyme product (Bines, Young, and Law 1989; Prokipek 1988). A 7-day acute oral toxicity study and a 91-day subchronic toxicity study in rats were done using the enzyme product. Using cheese made with the rDNA-derived chymosin, a 23-day oral toxicity test and a 91-day subchronic oral toxicity test were done in rats. No significant effects were indicated in any of the tests. In addition, studies of the enzyme's allergenicity, mutagenicity, and cell toxicity were performed. These tests were made in part because the enzyme is prepared as a dry product, not due to concern over rDNA technology involvement. In addition to the toxicity studies with the enzyme and cheese, extensive tests were conducted and submitted demonstrating the nonpathogenicity of the production organism. No Shiga-like toxin evaluation was conducted. These various investigations indicated the product had no safety concerns. The production containment procedures outlined in the environmental assessment for the recombinant host met Good Industrial Large Scale Practice (GILSP) guidelines (Teso 1986).

Chymosin Derived from *Aspergillus niger* var. *awamori*

In the most recent chymosin petition, submitted by Genencor, Inc. (1989) and accepted for filing by the FDA, the genetic information for prochymosin was cloned from the stomach cells of a preruminant calf into an *Aspergillus niger* var. *awamori* strain. The final plasmid contained a portion of the pBR322 plasmid vector, the calf prochymosin coding sequence, a *Neurospora crassa* pyr4 gene as a screen marker, and a host-specific glucoamylase promoter, genomic coding region, and terminator. The cDNA-encoding prochymosin B was fused in-frame immediately following the codon for the last amino acid of the glucoamylase gene. The mature chymosin is autocatalytically released from the glucoamylase–prochymosin fusion protein after secretion (Ward et al. 1990).

Toxicological examinations submitted included a 90-day subchronic toxicity study in rats as well as investigations of the allergenicity, mutagenicity, and cell toxicity potential of the product. Antibacterial and mycotoxin activity of the product were also investigated and found to be negative. Concerns for worker safety were addressed by studies of the potential for the product to cause dermal and eye irritation. In addition, extensive tests were conducted and submitted demonstrating the nonpathogenicity of the chymosin production organism. These various investigations indicated that the product had no safety concerns. The containment procedures described in the environmental assessment for the *Aspergillus* host met GILSP guidelines.

Bacillus stearothermophilus Maltogenic Amylase
Derived from *Bacillus subtilis*

Novo Laboratories has submitted and had accepted for filing a petition requesting affirmation of GRAS status for an exomaltohydrolase from *B. stearothermophilus* produced in *B. subtilis* (Novo Laboratories 1990). This enzyme is a newly developed enzyme that had not been previously commercialized. The gene for the amylase was cloned using the vector pUB110 into a nonpathogenic, nontoxicogenic, asporogenic *B. subtilis* host (Diderichsen and Christiansen 1988). The final plasmid in the production host was composed of a promoter and amylase coding region originating from *B. stearothermophilus,* a portion of the pUB110 plasmid including the kanamycin and phleomycin resistance markers, and approximately 150 base pairs of the DNA that may have arisen from the *E. coli* K-12 strain used as an intermediate cloning host.

Toxicological information contained in the GRAS petition included a 28-day subacute oral toxicity study and a 90-day subchronic oral toxicity study in rats, and a 21-day subchronic toxicity study in dogs. Additional tests included sensitization, mutagenicity, inhalation toxicity, and dermal and eye irritation assessments. Again, no safety concerns were uncovered during these investigations (Andersen et al. 1987).

DISCUSSION

This chapter summarizes the progress that has been made in the last decade with regard to safety evaluation and regulatory approval for the first food ingredients from genetically engineered organisms. In the 1980s we saw the development of rDNA-derived enzymes to a commercial level, the birth of appropriate strategies for evaluating their safety, and finally, the first affirmation of GRAS status by the FDA for one of these enzymes.

What will the 1990s hold for the new products of modern biotechnology? It is obvious from recent actions that the FDA will be even more active in this area. The FDA has announced that the agency is developing guidelines for the scientific review of food products developed through biotechnology (*Food Chemical News* 1990a); in fact, the first set of these guidelines from the Division of Food Chemistry and Technology has been issued (*Food Chemical News* 1990b). Additional guidelines focusing on microbiological, toxicological, and environmental assessment issues are expected shortly.

On April 2, 1990, the FDA announced its proposal to create an Office of Biotechnology in the Immediate Office of the Commissioner. The general goal of the office is "to enable [the] FDA to meet the new challenges presented by advances in the area of biotechnology." Some specific responsibilities of this office will be to:

Advise and assist the Commissioner and other key officials on scientific issues which have an impact on biotechnology policy, direction and long-range goals . . . [represent] the Agency on biotechnology matters to other government agencies, State and local governments, industry, academia, consumer organizations, Congress, national and international organizations, and the scientific community, [and to provide] leadership to Agency components in the identification, recruitment, and retention of top level scientists to fill vacancies for key Agency biotechnology positions. (Food and Drug Administration 1990b)

The agency has also established two "cooperative" biotechnology research facilities—the National Center for Toxicological Research and the National Laboratory for Food Safety and Technology, a cooperative effort with the Illinois Institute of Technology (IIT), the IIT Research Institute, and the University of Illinois. At these facilities, the FDA plans to bring together experts from academia, industry, and the government to promote its efforts to become actively involved in rDNA technology research (*Food and Chemical News* 1988).

In 1986, the Office of Science and Technology Policy (1986) and the participating federal agencies adopted a regulatory approach that was, to the extent permitted, to apply the existing laws and regulations to products derived from rDNA technology. Furthermore, the FDA stated that it would not propose new procedures or requirements for food regulation; the current regulations were sufficiently comprehensive to apply to products involving rDNA technology (Office of Science and Technology Policy 1984). To date, this approach has worked well for the regulation of rDNA-derived enzymes, simple food products. The capability of the enzyme industry to use this new technology and demonstrate product safety along with the agency's capability to regulate these products has now been established. What lies ahead are the challenges that will present themselves in the evaluation of safety and regulation of new, whole foods and food mixtures produced through modern biotechnology.

A whole food has never been the subject of a GRAS petition. All whole foods in commerce today were approved based on common, safe use prior to 1958; thus, there is no precedent for affirmation of GRAS status based on scientific procedures for a food such as a tomato. As evidenced earlier, the safety of food products such as enzymes is generally assessed in part by feeding animals the substance in amounts greatly exceeding the anticipated human daily intake. Whole foods and basic constituents of the human diet, such as starch, cannot be fed to animal models at levels in excess of average daily consumption without severely disrupting the nutritional balance of the animal's diet. Therefore, the anticipated use of genetic engineering to generate improved whole foods presents a challenge to food safety evaluation, not because genetic engineering will be employed, but because of the lack of accepted methods for safety evaluation of the new foods.

In February 1988, members of the food industry formed the International Food Biotechnology Council (IFBC), with the objective of "identifying issues and assembling a set of scientific criteria to evaluate and assure the safety of food and food ingredients derived from plants and microorganisms resulting from the application of biotechnology" (International Food Biotechnology Council 1990). The work product from this group is a report entitled "Biotechnologies and Food: Assuring the Safety of Foods Produced by Genetic Modification," released in final form in June 1990, which proposes a decision-tree approach for evaluating food products that fall in one of three categories: food and food ingredients derived from microorganisms; single chemicals and simple mixtures; and whole foods and other complex mixtures. The report proposes that the regulation of genetically modified food plants and microorganisms continues to be patterned on existing law and practice. The IFBC document also encourages that FDA affirm the practice of independent GRAS determinations by industry with respect to specified types of biotechnology-derived food products and that the agency establish an informal procedure by which industry can inform the agency of such independent determinations.

The safety evaluation and regulation of rDNA-derived products for the food industry are also progressing in other countries. Enzymes from rDNA organisms have been accepted for use in food by other nations, including Australia, Canada, and various members of the European Community. All six of the enzymes discussed in this chapter were agenda items at the thirty-seventh meeting (June 5–14, 1990) of the Joint Food and Agriculture Organization of the United Nations/World Health Organization Expert Committee on Food Additives in Geneva. The Committee evaluated information on use, identity, purity, specifications, and toxicological properties of the enzymes. All six enzyme products were granted "acceptable daily intake (ADI) not specified" status, meaning that, "on the basis of available data, the total daily intake of the substance arising from its use at the levels necessary to achieve the desired effect and from its acceptable background in food, does not, in the opinion of the Committee, represent a hazard to health." For these reasons, an ADI expressed in numerical form was not deemed necessary.

CONCLUSIONS

The new techniques of genetic engineering have opened the door for numerous possibilities for change in our food supply. From the previous discussions, it is apparent that the assessment of the safety of these biotechnology-derived food ingredients is a progressive and challenging field. With the cooperative efforts of the governmental agencies and industry we have and will continue to see the commercialization of new products from biotechnology that will provide safer, less expensive, purer, and more consistent food ingredients for human consumption.

References

Andersen, Jarl R., Borge K. Diderichsen, Rolf K. Hjortkjaer, Anne S. de Boer, James Bootman, Heather West, and Roger Ashby. 1987. Determining the safety of maltogenic amylase produced by rDNA technology. *J. Food Prot.* **50**:521–526.

Berkowitz, D., and J. Maryanski. 1989. "Implications of Biotechnology on International Food Standards and Codes of Practice." Paper presented at the Joint FAO/WHO Foods Standards Programme, Codex Alimentarious Commission, Eighteenth Session, 3–12 July 1989, Geneva.

Bines, Valerie E., Paul Young, and Barry A. Law. 1989. Comparison of cheddar cheese made with a recombinant calf chymosin and with standard calf rennet. *J. Dairy Res.* **56**:657–664.

CPC International, Inc. 1983. Filing of a petition for GRAS status (petition No. GRASP 3G0284). *Fed. Reg.* **48**:43,096.

CPC International, Inc. 1986. Filing of a petition for affirmation of GRAS status (petition No. GRASP 4G0293). *Fed. Reg.* **51**:10,571.

Diderichsen, Borge, and Lars Christiansen. 1988. Cloning of a maltogenic alpha-amylase from *Bacillus stearothermophilus*. *FEMS Microbiol. Lett.* **56**:53–60.

Enzyme Bio-Systems Ltd. 1988. Filing of a petition for affirmation of GRAS status (petition No. GRASP 7G0328). *Fed. Reg.* **53**:16,191.

Federal Food, Drug, and Cosmetic Act. 1982. 21 U.S. Code §§301–392.

Food Chemical News. October 17, 1988, p. 50.

Food Chemical News. 1990a, March 5, p. 3.

Food Chemical News. 1990b, May 28, p. 52.

Food and Drug Administration. 1982. Appendix II. In *Guidelines for Subchronic Oral Toxicity Studies, Toxicological Principles for the Safety Assessment of Direct Food Additives and Color Additives Used in Food,* pp 19–29. Reston, VA: Food and Drug Administration.

Food and Drug Administration. 1985. National environmental policy act; policies and procedures, final rule. *Fed. Reg.* **50**:16,636–16,669.

Food and Drug Administration. 1990a. Direct food substances affirmed as generally recognized as safe; chymosin enzyme preparation derived from *Escherichia coli* K-12. *Fed. Reg.* **55**:10,932–10,935.

Food and Drug Administration. 1990b. Statement of organization, functions, and delegations of authority. *Fed. Reg.* **55**:12,283–12,284.

Genencor, Inc. 1989. Filing of a petition for GRAS status (petition No. GRASP 9G0352). *Fed. Reg.* **54**:40,910–40,911.

Gibbs, Jeffrey N., and Jonathan S. Kahan. 1986. Federal regulation of food and food additive biotechnology. *Admin. Law Rev.* **38**:1–32.

Gist-Brocades, Inc. 1989. Filing of a petition for GRAS status (petition No. GRASP 9G0349). *Fed Reg.* **54**:20,203.

International Food Biotechnology Council. 1990. Biotechnologies and food: Assuring the safety of foods produced by genetic modification. *Reg. Toxicol. Pharmacol.* **12**(3):S1-S196.

Konowalchuk, J., J. I. Speirs, and S. Stavric. 1977. Vero response to a cytotoxin of *Escherichia coli. Infect. Immunol.* **18**:775-779.

MacKenzie, K. M., S. R. W. Petsel, R. H. Weltman, and N. W. Zeman. 1989*a*. Subchronic toxicity studies in dogs and in *in utero*-exposed rats fed diets containing *Bacillus megaterium* amylase derived from a recombinant DNA organism. *Food Chem. Toxic.* **27**:301-305.

MacKenzie, K. M., S. R. W. Petsel, R. H. Weltman, and N. W. Zeman. 1989*b*. Subchronic toxicity studies in dogs and in *in utero*-exposed rats fed diets containing *Bacillus stearothermophilus* alpha-amylase from a natural or recombinant DNA host. *Food Chem. Toxic.* **27**:599-606.

McNamara, Stephen H. 1987. FDA regulation of food substances produced by new techniques of biotechnology. *J. Food Drug Cosmetic Law* **42**:50-64.

Meisel, H. 1987. Charakterisierung von gentechnologisch gewonnenen labpraparaten im vergleich zu kalberlab 1, material and methoden. *Milchwissenschaft* **42**:787-789.

Meisel, H. 1988. Charakterisierung von gentechnologisch gewonnenen labpraparaten im vergleich zu kalberlab 2, ergebnisse. *Milchwissenschaft* **42**:71-75.

Metz, Raymond J., Larry N. Allen, Tin M. Cao, and Nancy W. Zeman. 1988. Nucleotide sequence of an amylase gene from *Bacillus megaterium. Nucleic Acids Res.* **16**:5203.

Miller, Sanford A., and Susan Thompson. 1985. "Regulating Applications of Biotechnology in the Food Processing Industry." Paper presented at the Regulatory Affairs Professionals Society Educational Seminar, Biotechnology Update 1985, 17 October 1985, San Francisco.

National Academy of Sciences/National Research Council, Food and Nutrition Board, Committee on Codex Specifications. 1981. Enzyme preparations. In *Food Chemicals Codex,* 3rd ed. pp. 107-110. Washington, D.C.: National Academy Press.

National Institutes of Health. 1986. Guidelines for research involving recombinant DNA molecules. *Fed. Reg.* **51**:16,958-16,985.

Novo Laboratories, Inc. 1990. Filing of a petition for GRAS status (petition No. GRASP 7G0326). *Fed. Reg.* **55**:9772-9773.

Office of Science and Technology Policy. 1984. Proposal for a coordinated framework for regulation of biotechnology. *Fed. Reg.* **49**:50,856-50,907.

Office of Science and Technology Policy. 1986. Coordinated framework for regulation of biotechnology. *Fed. Reg.* **51**:23,302-23,350.

Pfizer Central Research, Pfizer, Inc. 1988. Filing of a petition for affirmation of GRAS status (petition No. GRASP 8G0337). *Fed. Reg.* **53**:3792.

Prokipek, D. 1988. Herstellung von edamer und tilsiter kase mit gentechnologisch aus *K. lactis* gewonnenem rinder-chymosin. *Kieler Milchwirtsch. Forschungs.* **40**:43-52.

Teso, B. 1986. Recombinant DNA: From the lab to large-scale use, first steps toward international guidelines. *OECD Observer,* pp. 17–22.

Van den Berg, Johan A., Kees J. van der Laken, Albert J. J. van Ooyen, Ton C. H. M. Renniers, Krijn Rietveld, Albert Schaap, Anthony J. Brake, Robert J. Bishop, Kathleen Schultz, Donna Moyer, Michael Richman, and Jeffrey R. Shuster. 1990. *Kluveromyces* as a host for heterologous gene expression: Expression and secretion of prochymosin. *Bio/Technology* **8**:135–139.

Ward, Michael, Lori J. Wilson, Katherine H. Kodama, Michael W. Rey, and Randy M. Berka. 1990. Improved production of chymosin in *Aspergillus* by expression as a glucoamylase-chymosin fusion. *Bio/Technology* **8**:435–440.

Zeman, Nancy W., and Jan M. McCrea. 1985. Alpha-amylase production using a recombinant DNA organism. *Cereal Foods World* **30**:777–779.

5

Toward the Genetic Engineering of Disease Resistance in Plants: The Transfer of Pea Genes to Potatoes

Lee A. Hadwiger

The heightening concern of consumers about the pesticide residues of food has resulted in the loss of or requirement for reregistration of some of the mainstay pesticides used in agriculture (Richardson 1989). As this process continues, the need for employing any and all *natural* processes that can contribute to plant protection becomes paramount. The abundance and low cost of foods to date has contributed to the low level of funding for plant research. The plant scientist is now facing an urgent demand for natural plant protection without an extensive backlog of supportive basic research. That is, we are expected to replace the rare, effective chemicals derived from millions of synthetically generated compounds with natural compounds (Bell 1981; Bailey and Mansfield 1982) that are painstakingly derived from natural defense responses in plants. Further, these compounds must also be effective when applied externally, be scrutinized for safety concerns, be reasonable in price, and be applicable to the existing agricultural practices.

These are large orders, and in many cases such demands will not be answered. The strategy outlined in this chapter addresses one niche, the development of plant protection through genetic engineering, that we propose can be attainable. Many laboratories have contributed to the database on mechanisms of disease resistance (Dixon and Lamb 1990) and technology for plant transformation (Klee, Horsch, and Rogers 1987; An et al. 1985) from which we derive a narrow strategy for transferring pea genes to potatoes in an attempt to enhance the potato's resistance to disease. Plant breeders have demonstrated through the twentieth century that the most environmentally sound and "food-safe" mode of plant protection can be developed by redistributing (and rearranging) genes for disease resistance within a species to their commercially desirable counterpart cultivars. For many plants the advent of genetic engineering affords the opportunity for adding disease-resistance genes from *other* species. This breakthrough is phenomenal because there is the potential for *any* plant to provide beneficial genes to a given commercial crop.

NONHOST RESISTANCE

In general, all plants resist most of the microorganisms with which they come in contact (Yarwood 1973). Certain microbes have presumably through evolution acquired the ability to infect and reproduce on certain plant species, but yet are unsuccessful pathogens of other plants. These "other" plants that inhibit these pathogens are said to process "nonhost resistance" (Heath 1987). This nonhost resistance has not been characterized in terms of conventional Mendelian genetics, since the interspecies crosses needed to follow the distribution of genes in the progeny are rarely successful. However, the nonhost resistance response that is at the heart of the disease resistance process is associated with and probably requires the activation of multiple genes (Wagoner, Loschke, and Hadwiger, 1982), each of which can be cloned, characterized, and eventually utilized in genetic engineering schemes.

Currently one cannot readily discern with of the multiple genes in this response are the major contributors to disease resistance. However, it has been observed that treatments that prevent, alter, or delay this multiple gene response usually render the host susceptible to a much wider array of plant pathogenic organisms (Fernandez and Heath 1989; Hadwiger and Wagoner 1983a; Teasdale et al. 1974). That is, RNA or protein synthesis inhibitors can inhibit the response and break disease resistance. Additionally, in peas, treatments (Hadwiger and Beckman 1980; Kendra, Christian, and Hadwiger 1989) that are capable of enhancing the synthesis of these disease-resistance response proteins in advance of a challenge by an authentic pea pathogen, render the normally susceptible tissue resistant. Such tricks to attain resistance are compatible with a general principle of plant pathology, namely that disease resistance depends on the speed at which the plant response is generated. When the response is *pregenerated,* it apparently is sufficiently intense to maintain resistance even against a true pathogen (Kuc and Preisig 1984).

MENDELIAN RESISTANCE

Also, there is evidence that the Mendelian traits for disease resistance (Flor 1971), which the plant breeder crosses into commercial cultivars, are in some way instrumental in enhancing the activity of some of the same response genes active in nonhost resistance (Wagoner, Loschke, and Hadwiger 1982; Daniels et al. 1986). Therefore, disease resistance may often be a matter of regulating a more or less general resistance response rather than one of synthesizing a "silver bullet" directed against the pathogen. In this chapter I will integrate our past and current research results characterizing "nonhost" and "Mendelian" disease-resistance genes in peas, and the problems and prospects for their transfer to protect other crops such as potatoes that can be successfully transformed with foreign genes.

IDENTIFICATION OF DISEASE-RESISTANCE RESPONSE GENES

The first research objective is to identify (Riggleman, Fristensky, and Hadwiger 1985) those genes that are most active as the plant tissue is in the process of resisting a pathogen. We chose a root-rotting fungus, normally a pathogen of beans, that is totally resisted by pea tissue to activate the resistance response genes of peas.

Identification of these response genes is possible via their initial transcription products (mRNAs) (Hadwiger and Wagoner 1983b). The relative accumulation of these mRNAs can be tentatively evaluated by comparing in vitro translation products of mRNA from unchallenged healthy tissue with those of challenged, resisting tissue. The electrophoretic separation of these in vitro products provides information on their approximate molecular weights and isoelectric points (Wagoner, Loschke, and Hadwiger 1982). Further, the technology is now available for obtaining partial amino acid sequence analysis of the small quantities of proteins present in a two-dimensional gel separation. The partial amino acid sequence often predicts the appropriate nucleic acid sequence of the mRNA or original DNA sequence from which it was derived. A short nucleic acid probe can then be chemically synthesized that will recognize its own homologous nucleic acid(s) that are extractable from the plant tissue. This "cold" nucleic acid oligomer can also be generated into a highly radioactive probe that will hybridize with a homologous sequence of DNA in a genomic library or a cDNA library. These genes cloned from either library (Chiang and Hadwiger 1990; Fristensky, Horovitz, and Hadwiger 1988) can be sequenced to provide more complete information on the mRNA and the DNA sequences within the structural gene (including introns) and on the 3' and 5' regions of DNA adjacent to the structural gene that control the activation of the genes.

The disproportionate accumulation of certain mRNA species can be used directly as a basis to develop cDNA libraries. In this case the segments of plant DNA cloned into *E. coli,* which represent the most active genes, must be sorted out with radioactive probes representing RNA from the healthy control and the challenged resisting plants via plus-minus hybridization (Riggleman, Fristensky, and Hadwiger 1985). That is if the cloned DNA from a given clone hybridizes much more intensely with the probe representing all of the RNA of the resisting tissue than with that of the control, then it probably represents a gene actively induced.

Riggleman, Fristensky, and Hadwiger (1985) pursued the cloning of disease resistance clones in peas by constructing a cDNA library from mRNA isolated from pea endocarp tissue at the peak of disease-resistance expression. *E. coli* transformed with plasmids carying individual pea genes were allowed to grow to colonies 2 mm in diameter. Bacterial imprints of these colonies pressed onto a set of two round sheets of nitrocellulose were lysed and baked, which impregnates and exposes its DNA including that of the plasmid gene on this membrane surface. The first sheet is hybridized with uniformed preparations of a ^{32}P-labeled cDNA probe made with mRNA from untreated tissue, and the second sheet is hybridized with labeled mRNA from tissue actively resisting a bean pathogen. Colonies that bound greater amounts of induced probe than noninduced probe were selected and evaluated further for their role in the resistance response.

EVALUATION OF DISEASE-RESISTANCE
RESPONSE GENES

Once a quantity of clones representative of the actively induced genes is obtained, the clones can be further screened by evaluating if their expression in plants peaks in synchrony with the peak period of resistance (Fristensky et al. 1985) (Fig. 5-1). Unfortunately, the function of such genes often remains unknown and their expression is only *correlated* with disease resistance. Some functional information can be derived by examining the predicted amino acid sequences, which in turn suggests folding and helical structure in addition to identifying hydrophobic or hydrophilic regions of the protein product. Figure 5-2 indicates the similarities in hydrophobicity peaks among plant response genes found in potato, pea, and parsley. Obviously, these gene products also share high homologies in their amino acid sequences predicted from DNA sequence analyses (Matton and Brisson 1989). The amino acid composition of certain regions can determine if portions of the protein are capable of spanning membranes. Also, certain

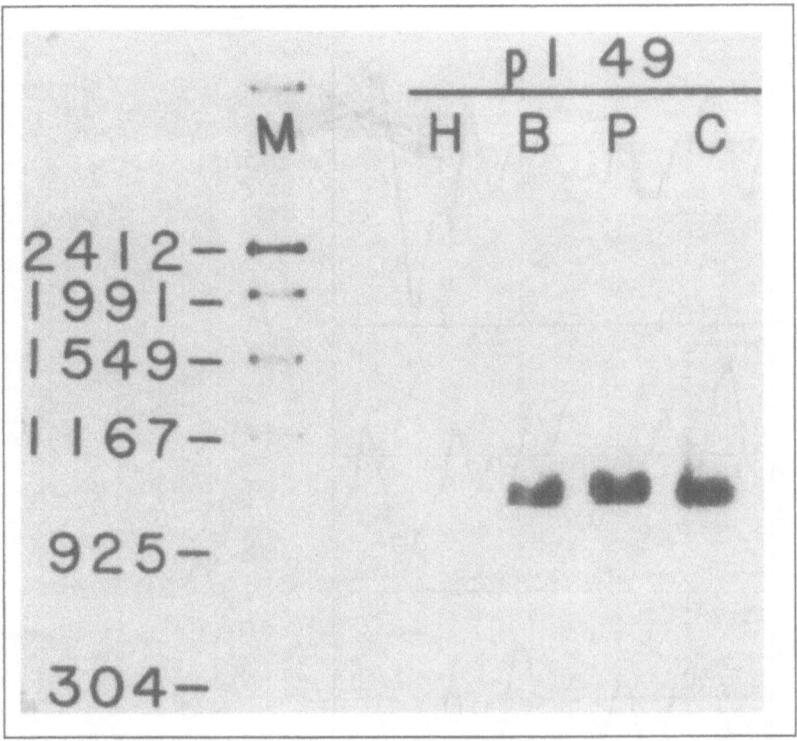

FIGURE 5-1.   A northern analysis of pea endocarp RNA probed with radioactively labeled nucleic acid homologous with gene 49, a disease response resistance gene. Healthy tissue (H) shows no accumulation; however, when the endocarp is challenged by a bean pathogen (B), a pea pathogen (P), or the elicitor chitosan (C), the gene transcript accumulates.   (From Fristensky et al., 1985, by permission of the copyright owner)

amino acid sequences are associated with nuclear transport, transfer through endoplasmic reticula, and so on (Newport and Forbes 1987; Hadwiger 1988; Silver 1991). Comparisons of both the nucleic acid and protein sequences with those already entered into computer banks can indicate if homologies exist with proteins or genes of defined function. At present many of the genes functioning in the disease resistance response cannot be associated with a known functional counterpart.

ROLE OF RESPONSE GENES WITH
KNOWN FUNCTIONS

Conversely, it has been possible to start with a definable functional entity of the host–parasite interaction and work toward isolating the gene coding

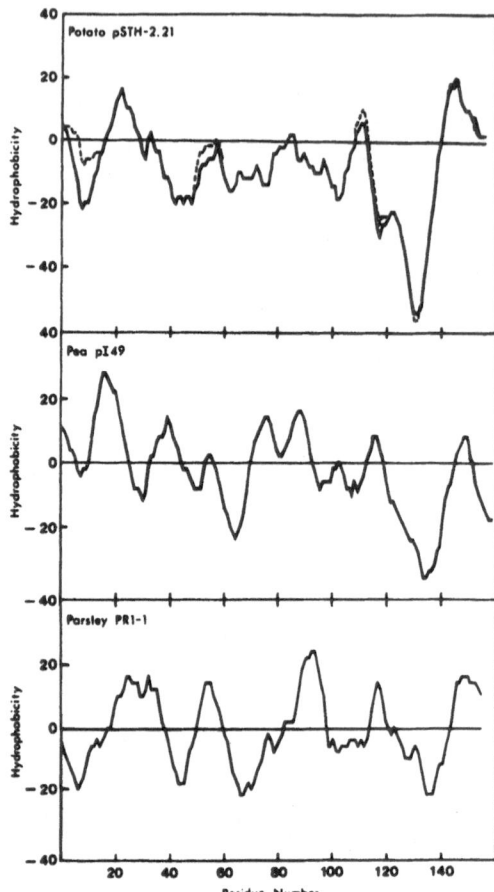

FIGURE 5-2.   Hydrophobicity curves of the predicted amino acid sequences of response genes from potato, pea, and parsley.   (From Matton and Brisson 1989)

this function. For example, the two plant enzymes β-glucanase and chitinase, contribute to the plant's defense since they digest β-glucan and chitin, respectively, which are the major structural polymers of the cell walls of many plant pathogenic fungi (Mauch, Hadwiger, and Boller 1984; 1988). Chitinase also has lysozymic activity that may also be detrimental to bacterial pathogens. In the *pea–Fusarium solani* interaction the activity levels of these enzymes increase in pea tissue following challenges. Interestingly, the intensity of the increase is somewhat similar regardless of whether the challenge is by the bean pathogen (*F. solani* f. sp. *phaseoli*) or the pea pathogen (*F. solani* f. sp. *pisi*) (Fig. 5–3). However, there appears to be a requirement for these enzymes in the plant's resistance, because in the plant–enzyme and fungal-cell interaction there is a release of carbohydrate oligomers (Fig. 5–4), including segments of deacetylated chitin called chitosan (Kendra and

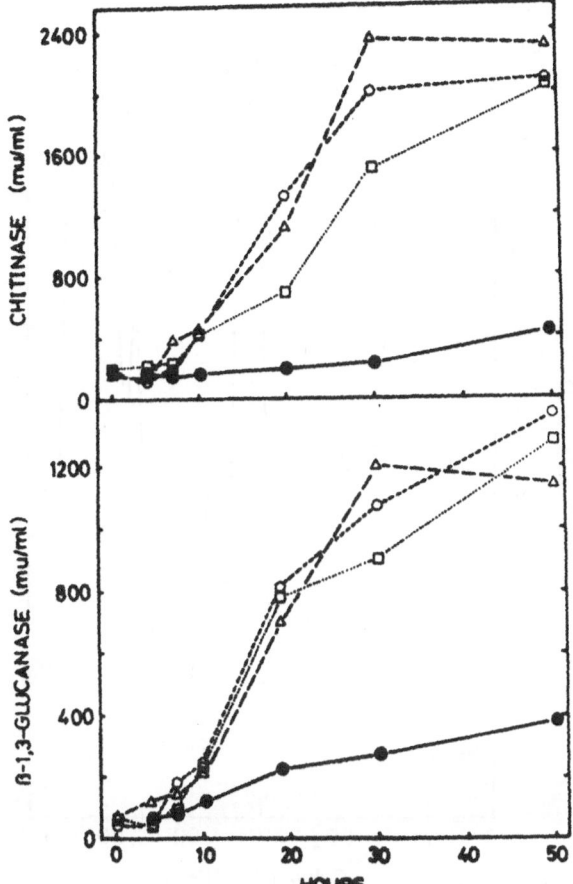

FIGURE 5-3. Induction of chitinase activity (top) and β-1,3-glucanase activity (bottom) in immature pea pods by a compatible pathogen (open triangle), by an incompatible pathogen (open circle), and by chitosan (open square). Controls (closed circle) received sterile water. (From Mauch, Hadwiger, and Boller 1984, by permission of the copyright owner)

Hadwiger 1984; Kendra, Christian, and Hadwiger 1989). Chitosan at 10–20 μg/ml effectively stops germination and the mycelial growth of both fungal pathogens. Chitosan is a basic, positively charged polymer that can also induce the same pattern of genes in peas that are induced by a challenging bean pathogen (resistance reaction). If applied prior to challenge by the pea pathogen, it can generate *induced resistance* in peas to a level adequate for the pea to resist the pea pathogen (Hadwiger and Beckman 1980). Because β-glucanase and chitinase can be purified to homogenicity, partial amino acid sequence data have been obtained that, as previously described,

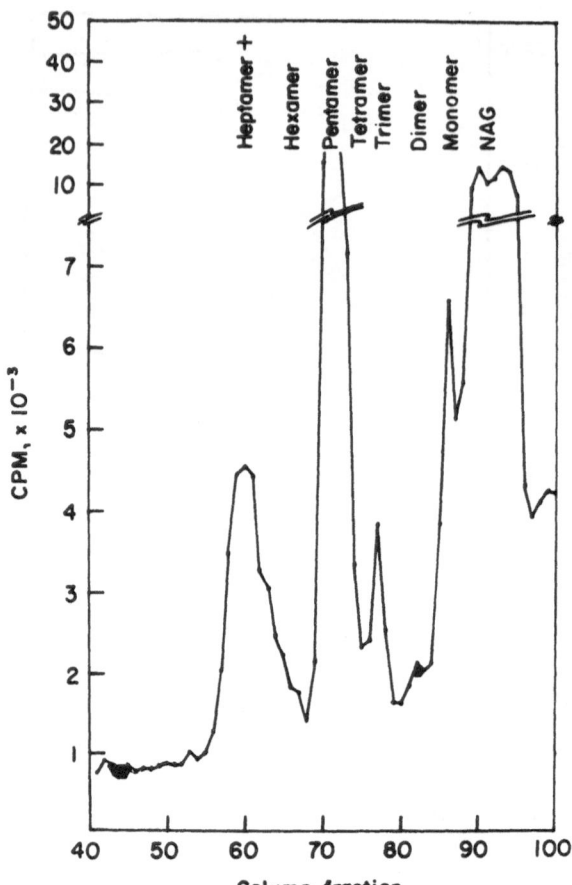

FIGURE 5-4.   An array of glucosamine polymers from fungal spores is released when the spore contacts living plant tissue or is incubated with purified chitinase and β-glucanase from pea tissue. The peaks eluted from a fractogel column were prelabeled in the fungus as it metabolized N-acetyl glucosamine-^{3}H. Biological functions of the heptamer peak (chitosanlike) include induction of a disease-resistance response in pea endocarp tissue and inhibition of the germination and growth of *Fusarium solani* spores.   (From Kendra, Christian, and Hadwiger, 1989, by permission of the copyright owner)

allow a nucleic acid probe to be constructed for screening cDNA and genomic libraries of cloned genes. cDNA expression libraries can also be screened via antisera developed against the β-glucanase and chitinase antigens.

In addition to these response genes being candidates for transfer of resistance between plant species, there are many single genes factors (the Mendelian traits mentioned earlier) that appear to exert an overall control over these response genes. These master genes often are associated with specific pathogens in gene-for-gene relationships (Flor 1971; Keen 1990; Lindgren et al. 1988). That is, when a given plant disease resistance trait (usually dominant) is present in an interaction with a given race of the pathogen that contains a corresponding avirulence trait (usually dominant), there is a pronounced enhancement and duration of expression of the response genes (slave genes) (Hadwiger, Chiang, and Horovitz 1991; Daniels et al. 1986) (Fig. 5-5). This master gene control is not functional if the host–parasite matchup does not include each of the corresponding dominant traits. At present it is not clear how the product of the plant's dominant trait functions (Bowles 1990; Keen 1990). Some correlation may exist between the master regulation by these traits and the master regulation of subsets of genes by oncogenes (Wessler and Hake 1990; Schmidt et al. 1990). The only clues available to date are that plants do produce proteins that recognize DNA sequences similar to those recognized by oncogene products (Singh et al. 1990). Also, a sequence has been found that is associated with a disease-resistance response gene that is identical to the sequence

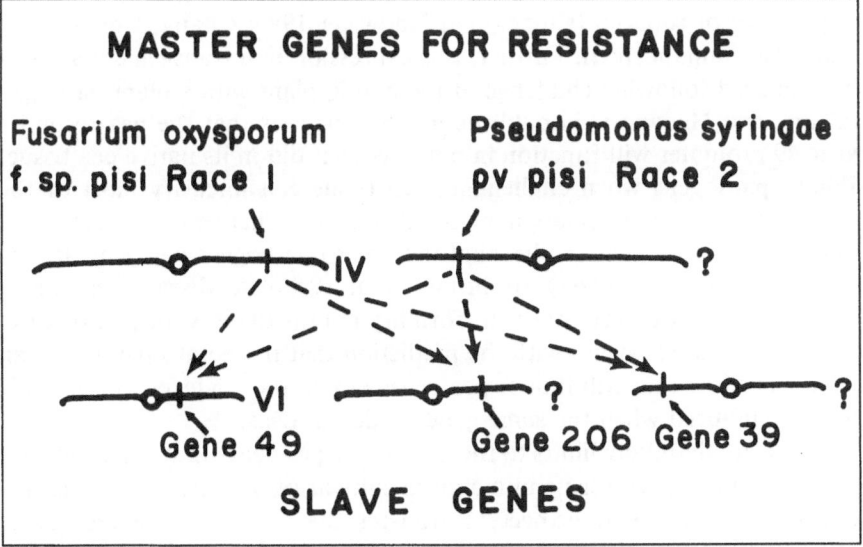

FIGURE 5-5.   Proposed hierachy of master genes representing single-dominant-factor determinants of race-specific resistance and slave genes representing cloned genes that become active. The slave genes are expressed both in race-specific and nonhost resistance. In one instance (Hadwiger et al. 1991), a pea line with a single dominant gene for resistance to *F. oxysporum* race 1 is located on chromosome IV and one of the response genes (slave gene) that becomes activated is on chromosome IV. The dotted lines suggest communications possibility by transacting factors between chromosomes. The presence of another single dominant trait for controlling resistance to a race of *Pseudomonas syringae* (Daniels et al. 1986) is also associated with induced increases in the slave genes.

recognized by an oncogene (Chiang and Hadwiger 1990). It is hoped that the cloning and sequence analysis of any of these plant traits, which we currently know only by conventional genetic linkage analysis, will provide further insights into how the master control functions. The availability of such master genes for transfer could reduce the number of slave genes needed for transferring resistance between species. Genome mapping efforts currently underway are likely to make such genes available in the future. Genetic manipulations such as overexpressing the diverse resistance response genes (slave genes) that appear to be the actual genes functioning in resistance in transgenic plants may be the next best technique to improve disease resistance. Thus, incremental increases in disease resistance in plants may be possible through such genetic engineering efforts without the requirement for cloned master genes. Our strategies have been aimed in this direction. The genomic clone of pea gene 49 (screened from a genomic library with the type of a probe previously mentioned) codes a protein of 17,000 molecular weight. This protein constitutes a major band of the total pea protein (Chiang and Hadwiger 1990) following fungal challenge. The 5' region of this genomic clone is capable of expressing a reporter gene such as the *chloramphenicol acetyl transferase* structural gene (CAT gene) in potatoes or tobacco (Chiang and Hadwiger 1990; Chang, Chiang, and Hadwiger, unpublished). Further, this expression in a transgenic plant can be enhanced following challenge of the potato plant with a plant pathogen (Chang and Hadwiger, unpublished). This suggests that the pea response gene 49 promoter will function in potatoes as it did in its native pea tissue. Potato pathogens when challenging pea tissue can intensify most of the measurable defense response in peas. Also the fact that pea tissue can resist most potato pathogens implies that the potato pathogens can generate the pea's response adequately to bring about complete disease resistance. Therefore, we are currently transforming potato tissue with pea disease-resistance response genes with the prediction that the potato pathogen can release elicitors that will induce pea genes in potatoes to levels comparable to those induced when the same gene resided in peas.

There are definitely limits to the number of pea genes that can be assembled via genetic engineering into a given cultivar of potatoes. Also it is currently unreasonable to expect to transfer the entire disease resistance-response entity of peas to potatoes. It is more reasonable to expect that a few transferred genes may be capable of augmenting the potato plant's *inherent* resistance. There are two reasons for why this resistance may occur. In our work with peas we have observed that the pea resists the growth of a highly virulent pathogen for a limited period, and employs the same defense genes as it does in resisting other pathogenic fungi; however, the pea response is short term in the presence of the virulent pathogen. Fortunately, the potato inherently possesses a response gene that closely resembles the pea gene 49 (Matton and Brisson 1989), but presumably it is controlled differently in a potato–potato pathogen interaction. Thus, the presence in potatoes of the additional pea gene 49 should augment the inherent defense response of the potato. Also, the added pea gene should function continuously throughout the challenge by a potato pathogen. Although the function of gene 49 has not been resolved, the observation that it can constitute up to 25% of the labeled protein of legumes (LeGuay et al. 1988) under induced conditions suggests some importance of this protein to the plant response. In the same vein the addition of pathogen-inducible plant genes, such as β-glucanase and chitinase from peas, should also contribute to the potato plant's response to help approach a threshold level of total response similar to that attainable in induced-resistance experiments.

CONCLUSIONS

There *are* molecular approaches available to augment the efforts of plant breeders to improve disease resistance in plants. This paper describes an approach used to identify and clone plant genes that are selectively activated as disease resistance is expressed. The diversity of other approaches will increase as more becomes known about the molecular nature of resistance. Also the genome mapping and sequencing (Wessler and Hake 1990) should help uncover additional disease resistance traits along with new understandings of how the master gene control over the induced slave genes functions. Many homologies exist within the PR (pathogenesis related) (Bowles 1990) and other response genes between plant species. Thus the genetic manipulations proposed in this paper involve only the enhancement of groups of plant proteins that are similar to those the plant already possesses. It is therefore unlikely that exotic or potentially harmful compounds will be introduced to alter the level of food safety we presently enjoy in cultivated crops.

References
An, G., B. D. Watson, S. Stachel, M. P. Gordon, and E. W. Nester 1985. New cloning vehicles for transformation of higher plants. *EMBO J.* **42**:277–284.
Bailey, J. A., and J. W. Mansfield. 1982. *Phytoalexins.* NY: Wiley.
Bell, A. A. 1981. Biochemical mechanisms of disease resistance. *Ann. Rev. Plant Physiol.* **32**:21–81.
Bowles, D. J. 1990. Defense related proteins in higher plants. *Ann. Rev. Biochem.* **59**:873–907.
Chiang, C. C., and L. A. Hadwiger. 1990. Cloning and characterization of a disease resistance response gene in pea induced by *Fusarium solani. Mol. Plant Microbe Interactions* **3**:75–87.

Daniels, C. H., B. W. Fristensky, W. Wagoner, and L. A. Hadwiger. 1986. Pea genes associated with non-host disease resistance to *Fusarium* are also active in race-specific disease resistance to *Pseudomonas*. *Plant Mol. Biol.* **8**:309–316.

Dixon, R. A., and C. J. Lamb. 1990. Molecular communications in interactions between plants and microbial pathogens. *Ann. Rev. Plant Physiol.* **41**:339–367.

Fernandez, M. R., and M. C. Heath. 1989. Interaction of the non-host French bean plant (*Phaseolus vulgaris*) with parasitic and saprophytic fungi. I. Fungal development on and in killed, untreated, heat-treated, or Blasticidin S treated leaves. *Can. J. Bot.* **67**:661–669.

Flor, H. H. 1971. Present status of the gene for gene concept. *Ann. Rev. Phytopathol.* **9**:275–296.

Fristensky, B., D. Horovitz, and L. A. Hadwiger. 1988. cDNA sequences for pea disease resistance response genes. *Plant Mol. Biol.* **11**:713–715.

Fristensky, B., R. C. Riggleman, W. Wagoner, and L. A. Hadwiger. 1985. Gene expression in susceptible and disease resistant interactions of peas induced with *Fusarium solani* pathogens and chitosan. *Physiol. Plant Pathol.* **27**:15–28.

Hadwiger, L. A. 1988. Possible role of nuclear structure in disease resistance in plants. *Phytopathology* **78**:1009–1014.

Hadwiger, L. A., and J. M. Beckman. 1980. Chitosan as a component of pea-*Fusarium solani* interactions. *Plant Physiol.* **66**:205–211.

Hadwiger, L. A., and W. Wagoner. 1983*a*. Effect of heat shock on the mRNA-directed disease resistance response of peas. *Plant Physiol.* **72**:553–556.

Hadwiger, L. A., and W. Wagoner. 1983*b*. Electrophoretic patterns of pea and *Fusarium solani* proteins synthesized *in vitro* which characterize the compatible and incompatible interactions. *Physiol. Plant Pathol.* **23**:153–162.

Hadwiger, L. A., C. C. Chiang, and D. Horovitz. 1991. Expression of disease resistance response genes in near isogenic pea cultivars following challenge by *Fusarium solani* race 1. *Physiol. Molec. Plant Pathol.* (in press).

Heath, M. C. 1987. Host vs. non-host resistance. In *Molecular Strategies for Crop Protection,* ed. C. J. Arntzen and C. Ryan, pp. 25–34. New York: Liss.

Keen, N. T. 1990. Gene-for-gene complimentarity in plant-pathogen interactions. *Ann. Rev. Genet.* **24**:447–463.

Kendra, D. F., and L. A. Hadwiger. 1984. Characterization of the smallest chitosan oligomer that is maximally antifungal to *Fusarium solani* and elicits pisatin formation in *Pisum sativum. Exp. Mycol.* **8**:276–281.

Kendra, D. F., D. A. Christian, and L. A. Hadwiger. 1989. Chitosan oligomers from *Fusarium solani*/pea interactions, chitinase/β-glucanase digestion and fungal wall chitin actively inhibit fungal growth and induce disease resistance. *Physiol. Mol. Plant Pathol.* **35**:215–230.

Klee, H. R. Horsch, and S. Rogers. 1987. *Agrobacterium*-mediacted plant transformation and its further applications to plant biology. *Ann. Rev. Plant Physiol.* **38**:467–486.

Kuc, J., and C. Preisig. 1984. Fungal regulation of disease resistance mechanisms in plants. *Mycologia* **76**:767–784.

LeGuay, J. J., M. Piecoup. J. Puckett, and J. P. Jouanneau. 1988. Common responses of cultured soybean cells to 2,4-D starvation and fungal elicitor treatment. *Plant Cell Rep.* **7:**19–22.

Lindgren, P. B., N. J. Panopoulos, B. J. Staskawicz, and D. Dahlbeck. 1988. Genes required for pathogenicity and hypersensitivity are conserved and interchangeable among pathovars of *Pseudomonas syringae. Mol. Gen. Genet.* **211:**499–506.

Matton, D. P., and N. Brisson. 1989. Cloning, expression, and sequence conservation of pathogenesis-related gene transcripts of potato. *Mol. Plant Microbe Interactions* **2:**325–331.

Mauch, F., L. A. Hadwiger, and T. Boller. 1984. Ethylene: Symptom, not signal for the induction of chitinase and β-1,3-glucanase in pea pods by pathogens and elicitors. *Plant Physiol.* **76:**607–611.

Mauch, F. C., L. A. Hadwiger, and T. Boller. 1988. Purification and characterization of two β-1,3-glucanase differentially regulated during development and in response to fungal infection. *Plant Physiol.* **87:**325–333.

Newport, J. W., and D. J. Forbes. 1987. The nucleus: Structure, function and dynamics. *Ann. Rev. Biochem.* **56:**535–566.

Richardson, L. 1989. Registration reality. *Agriculture Age* **33:**13–15.

Riggleman, R. C., B. Fristensky, and L. A. Hadwiger. 1985. The disease resistance response in pea is associated with increased levels of specific mRNAs. *Plant Mol. Biol.* **48:**81–86.

Schmidt, R. J., F. A. Burr, M. J. Aukerman, and B. Burr. 1990. Maize regulatory gene opaque-2 encodes a protein with a leucine-zipper motif that binds to zein DNA. *Proc. Natl. Acad. Sci. (USA)* **87:**46:50.

Silver, P. A. 1991. How proteins enter the nucleus. *Cell* **64:**489–497.

Singh, K., E. S. Dennis, J. G. Ellis, D. J. Llewellyn, J. G. Tokuhisa, J. A. Wahleithner, and W. J. Peacock. 1990. OCSBF-1 a maize Ocs enhancer binding factor: Isolation an expression during development. *Plant Cell* **2:**891–903.

Teasdale, J., D. Daniels, W. C. Davis, R. Eddy, Jr., and L. A. Hadwiger. 1974. Physiological and cytological similarities between disease resistance and cellular incompatibility responses. *Plant Physiol.* **54:**690–695.

Wagoner, W., D. C. Loschke, and L. A. Hadwiger. 1982. Two-dimensional electrophoresis analysis of *in vivo* and *in vitro* synthesis of proteins in peas inoculated with compatible and incompatible *Fusarium solani. Physiol. Plant Pathol.* **20:**99–107.

Wessler, S., and S. Hake. 1990. Maize harvest. *Plant Cell* **2:**495–499.

Yarwood, C. E. 1973. Some principles of plant pathology, II. *Phytopathology* **63:**1324–1325.

6

Automated System for Microbial Screening/Breeding

Toru Okuda and Hiroyoshi Tabuchi

New areas in biotechnology have progressed remarkably and are expected to produce new industries. Its methods have already been applied to a wide range of fields, such as agriculture, the chemical, food, energy and pharmaceutical industries. The key factors in biotechnology are gene manipulation, cell fusion, large-scale cell culture, and technology for bioreactors; pertinent industrial processes will require highly developed and precise equipment. The actual market scale of biotechnology has enlarged to more than 100 billion yen ($0.7 billion) in 1989 according to Nikkei Biotech (1989). Although an investigation by BIDEC (Bioindustry Development Center in the Japanese Association of Fermentation and Industry) reports that the industrial scale of biotechnology will grow up to 15,000 billion yen ($100 billion) per year by the twenty-first Century, most of the technology is now in the research and development stage. Rapid development is expected.

Further growth of biotechnology requires upgraded and sophisticated instruments and systems. Leading electronics industries have devised a wide variety of scientific instruments that incorporate basic electronics technology, image analysis, precise data processing, system control, and robotics.

First, an overview is presented of the kinds of instruments used in biotechnology and instruments used for DNA recombinant studies. The processes to be automated include incubation, disruption, separation, concentration, extraction, analysis, reaction, sterilization, prevention of contaminants, and storage. Since researchers have to handle trace amounts of biological substances, and need to analyze them at the molecular level, ac-

The figures in this chapter have been reproduced, with minor modifications, from Nukumi et al., 1987, by permission.

curacy is required; therefore, interfacing of processes is important and systematization is required.

Instruments for gene manipulation research include: incubators, deep freezers, centrifuges, spectrophotometers, gas chromatographs, mass spectrometers, nuclear magnetic resonance (NMR) spectrometers, and electron microscopes. With computer-aided automatic control, the denoted instruments have become highly accurate and laborsaving.

Once a target gene has been cloned and transferred into a microorganism, the gene needs to be expressed to produce the desired protein. Because the output per cell is constant, high-density fermentation is required for enhancing production efficiency. Sophisticated culture control technology is a necessity. However, scaleup fermentation is not always easy to handle because cell growth and substance production are influenced by a number of factors, and because biomass and product concentration are hard to measure during the fermentation. To solve the problem, accurate analysis of growth and metabolic activity of various organisms has been examined. For example, through examination of optimum control methods for the cultivation of recombinant yeasts or *Escherichia coli,* a high-density automatic control unit for fermentation has been developed by Hitachi (Fukujoji et al. 1987).

Further progress in biotechnology will be achieved by the incorporation of new methodologies and automated equipment. Thus, collaboration between researchers in different areas is essential for advancing fundamental work in biotechnology. However, adequate automated or sophisticated equipment is often not commercially available, and if it is developed for specific purposes, it is not known whether the process or research to be automated will be long-term. The process to be automated must therefore be defined before automated equipment is developed.

Two distinct robot systems that were developed in collaboration with engineers employed at leading electronics companies are described.

MICROBIAL SCREENING SYSTEM

Genetic engineering, or gene manipulation, has become an essential tool for the creation of innovative drugs. However, microbial screening has also been in the forefront of new drug discoveries, and new recent technologies and automated systems should be developed.

Screening for new antibiotics has a long history. Since the discovery of antibiotics such as penicillin and streptomycin, the pharmaceutical industries have been searching for new antibacterial substances from microorganisms. More than 5000 antibiotics have been discovered, mainly from actinomycetes, since the 1950s; this taxon has been a tremendous source of useful microbial metabolites such as penicillins and cephalosporins. The long and intensive search for antibiotics resulted in the discovery of important and effective antibiotic agents from microbial metabolites, such as β-lactams, aminoglycoside antibiotics, macrolides, tetracyclines, ansamycins, vancomycins, and chloramphenicol.

In the last decade the progress of pharmacology, biochemistry, and biotechnology has accelerated the understanding of serious human disease at the enzyme and molecular levels. A number of peptide hormones and other proteins have been found to play key factors in the control and regulation of cellular functions. Many opportunities exist for the development of new drugs from enzymes and peptides. Innovative drugs could be found through the random screening of microbial metabolites; this method has been improved, since assay systems have changed to identify more target-oriented objectives, such as enzyme inhibitors or protein–protein receptor antagonists or agonists.

The late Prof. H. Umezawa, the Institute of Microbial Chemistry, Japan, often stated that microorganisms had always measured up to his expectations. On the basis of his philosophy, he and his colleagues discovered a new aminopeptidase B inhibitor, bestatin (Fig. 6-1) from *Streptomyces olivoreticuli* (Umezawa et al. 1976). Bestatin clinically activated T cells and enhanced the activity of natural killer cells in patients with cancer. It is now used as a potent immunomodulator in the treatment of cancer. Immunotherapy is considered to be an important therapy in combination with surgery, radiation, and chemotherapy. Other than bestatin, Umezawa and his researchers found a number of potent enzyme inhibitors located on the cell surface, which would act as immunomodulators. They are esterasin from *Streptomyces lavendulae* (Umezawa et al. 1978), ebelactones from *Streptomyces* sp. (Umezawa et al. 1980) and arphamenines from *Chromobacterium violaceum* (Umezawa et al. 1983).

Cyclosporins (Fig. 6-2) from *Tolypocladium inflatum* (Dreyfuss et al. 1976) are cyclic peptides that were originally categorized as antifungal antibiotics, but are now recognized as highly potent immunosuppressive agents (Borel et al. 1977). For example, cyclosporin A characteristically inhibits

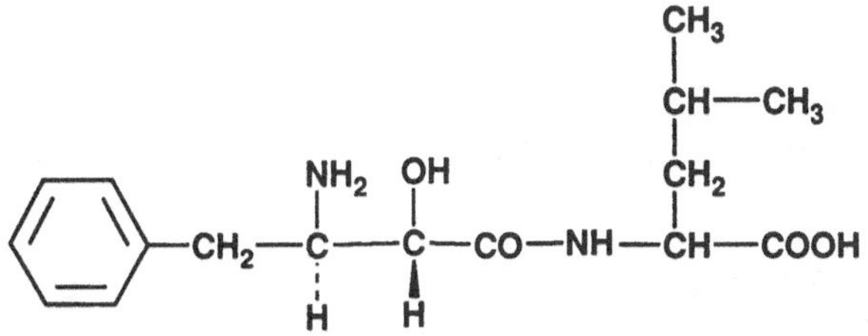

FIGURE 6-1.   Bestatin.

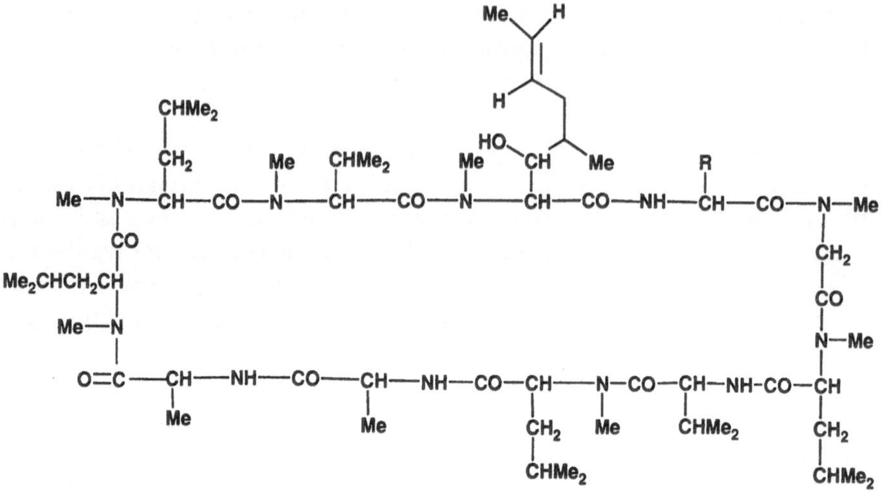

FIGURE 6-2.   Cyclosporins: A, R = Et; B, R = Me; C, R = CH(OH)Me; D, R = CHMe₂; G, R = (CH₂)₂Me.

the activation of T cells in the early immune response, which is used to prolong the survival of organ transplants.

FK-506 (Fig. 6–3), another immunosuppressive macrocyclic lactone produced by *Streptomyces tsukubaensis,* is more potent than cyclosporin A, because it attacks sites specific for the immunological response and strongly suppresses the production of interleukin 2 without cytotoxicity (Kino et al. 1987).

3-Hydroxy-3-methylglutaryl-coenzyme A reductase (HMG-CoA reductase) is responsible for the rate-limiting step of mammalian cholesterol bio-

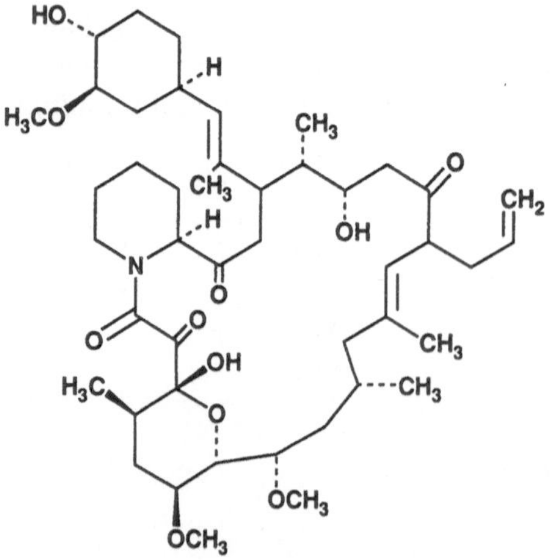

FIGURE 6-3.   FK-506.

synthesis. Endo et al. (1976; 1979) considered this reductase a promising target for the development of an antihypercholesterolemic drug; they discovered potent inhibitors, compactins, in *Penicillium* spp. and monacolins from *Monascus ruber*. Sankyo Company Ltd. launched a derivative of the compactins, Plavastatin, which is expected to have yearly sales of $3 million. Merck & Company, Inc. has also developed HMG-CoA reductase inhibitors, Simvastatin and Lovastatin (Fig. 6-4), that will compete with Sankyo's drug.

Cholecystokinin (CCK), a stimulant of gallbladder contraction and pancreatic enzyme secretion, and gastrin are peptide hormones that have identical terminal amino acid sequence; they are involved in the stimulation of gastric acid secretion. CCK occurs widely in the central nervous system, and it appears to be a neurotransmitter. Gastrin also exhibits trophic effects on target tissues. The receptor antagonists of CCK or gastrin are expected to be potential therapeutic agents in conditions such as pancreatitis, pancreatic carcinoma, anorexia, and irritable bowel syndrome.

Asperlicins were discovered as products of *Aspergillus alliaceus,* and the assay system was based on radioligand binding for rat pancreas and guinea

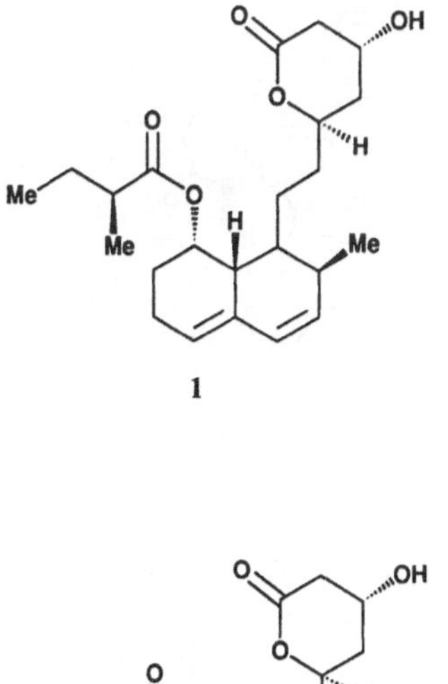

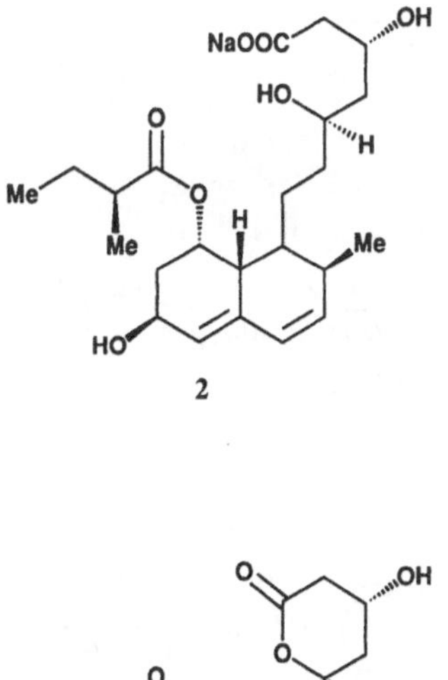

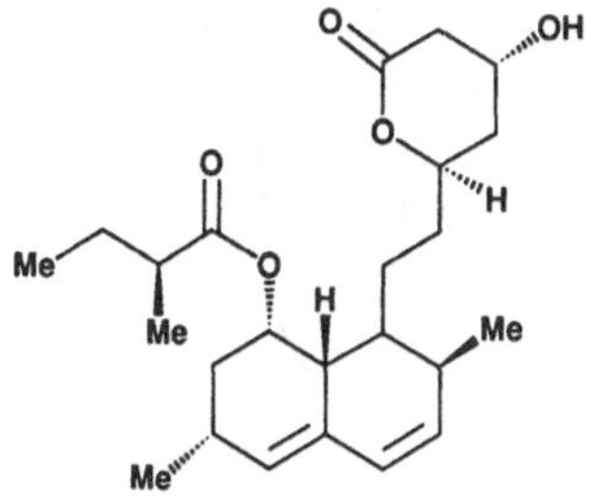

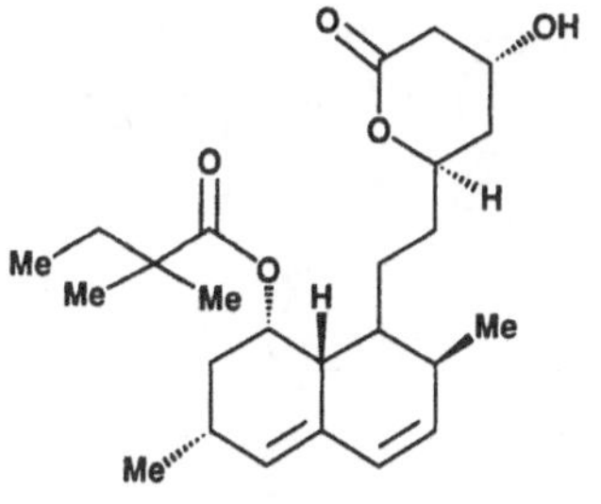

FIGURE 6-4.   HMG-CoA reductase inhibitors. (1) Mevastatin (Compactin) by Sankyo. (2) Plavastatin by Sankyo. (3) Lovastatin by Merck. (4) Simvastatin by Merck.

pig brain CCK receptors (Goetz et al. 1985). Although naturally occurring asperlicins did not possess the requisite potency to be used as therapeutic agents (IC50 of asperlicin is 1.4 μM for rat pancreas), several chemically modified derivatives (Fig. 6–5) exhibited enhanced activity due to their increased solubility in water. L-364,718 showed peripheral CCK selective activity with an IC50 of 8.0 $\times$ 10^{-9} μM for rat, and L-365,260 exhibited central CCK selectivity with an IC50 of 2.0 $\times$ 10^{-6} μM in guinea pig (Frei-dinger 1989).

Recent, promising compounds have characteristically been derived from a wide variety of taxa such as actinomycetes, fungi, and bacteria. Microorganisms can produce unique compounds beyond the ones synthesized by organic chemists, but development of them as therapeutic drugs may be restricted by their lack of selectivity, cell permeability, or stability. Experience in the field of medicinal chemistry, however, enhances the ability to create innovative drugs from compounds discovered in cultured broths of microorganisms. For instance, various important cephalosporin derivatives were launched from the weakly active natural cephalosporin C. Microbial screening in the beginning of the antibiotic era meant a process to discover antibacterial substances from actinomycetes. The present microbial screen-

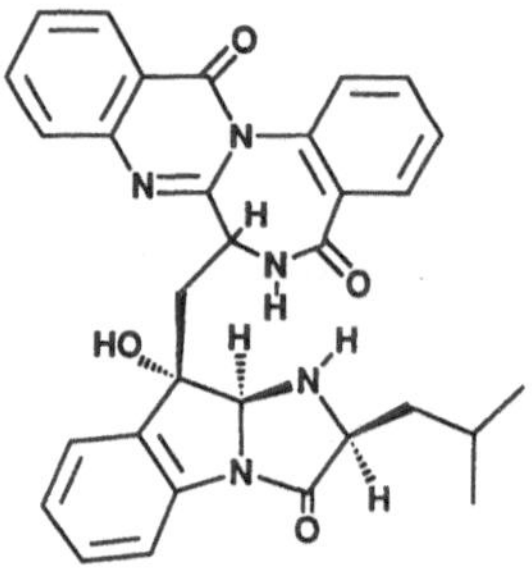

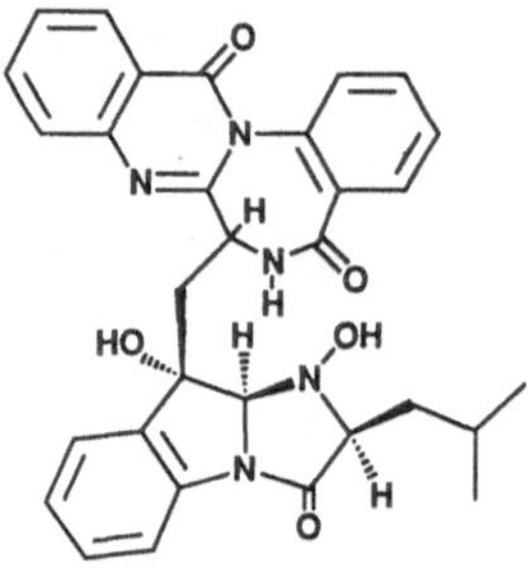

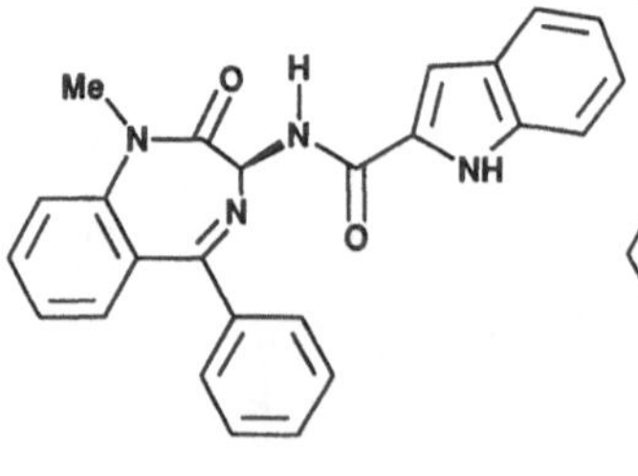

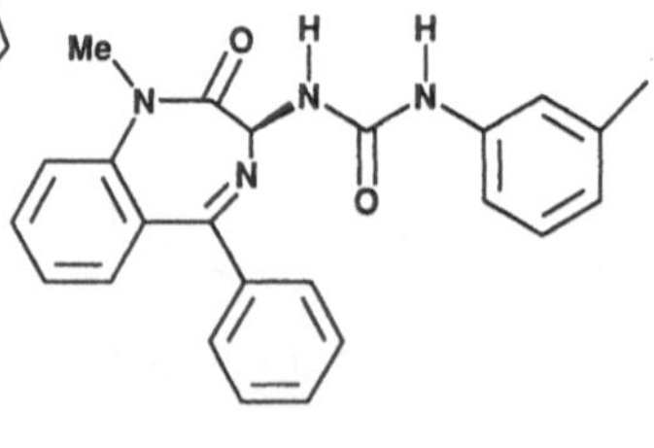

FIGURE 6-5.   CCK antagonists. (1) Asperlicin; (2) asperlicin B; (3) L-364,718; (4) L-365,260.

ing can be defined as a process to uncover relevant compounds from a wide range of microorganisms for innovative therapeutic agents.

Microbial screenings in the 1950s and 1960s were done on a small scale and were neither systematic nor intensive. For the most part, all aspects of the screening were done by a few researchers. Microorganisms would be isolated and cultured, then the cultured broth would be subjected to a specific antibacterial assay. Individual researchers may have even purified the active ingredient and elucidated its structure. All in all, this method proved to be less fruitful than anticipated for the discovery of promising compounds. In other fields, such as the music of the baroque era, an all-encompassing approach was also attempted in which one person acted as the composer, conductor, interpreter, and sometimes even manager. It is difficult to conceive of an individual carrying out a performance of large orchestral works of the late romantic or contemporary periods, such as those by Wagner, Mahler, or Richard Strauss. In order to succeed in the contemporary microbial screening, a collaboration of taxonomists, microbiologists, pharmacologists, biochemists, organic chemists, and physicochemists is therefore required. Pharmacologists and biochemists are responsible for setting up simple, unique, and selective assay systems that can detect minor but specific compounds in the cultured broths. The systems should avoid interference from nonselective compounds, such as high molecular protease or salts contained in the broths. Taxonomists or microbiologists should isolate a large number of microorganisms, which are taxonomically and physiologically different from each other. Relevant scientists should be knowledgeable about conventional morphology, and also about biosynthesis and the production of secondary metabolites. Sufficient productivity and reproducibility of the producing microorganisms must be maintained. Chemists have to isolate and characterize minute amounts of active compounds in small portions of cultured broth. Known compounds without promise should be excluded at an early stage of screening. Finally, managers should keep in mind that microbial screening is a long-term process.

Contemporary microbial screening is a large project, in which researchers from various fields are involved in many assay systems that require repetitive tasks. Therefore, application of an automated system to a certain process will reduce the amount of routine work done by the researchers and will make the operation of the screening more efficient.

In order to define which part of the system can be automated, the following process analysis is possible. Generally speaking, microbial screening contains microbiological, biochemical, and chemical aspects. Integration of research in these areas is rather complicated, but the whole system (Fig. 6-6) illustrates the necessity of close collaboration among the activities.

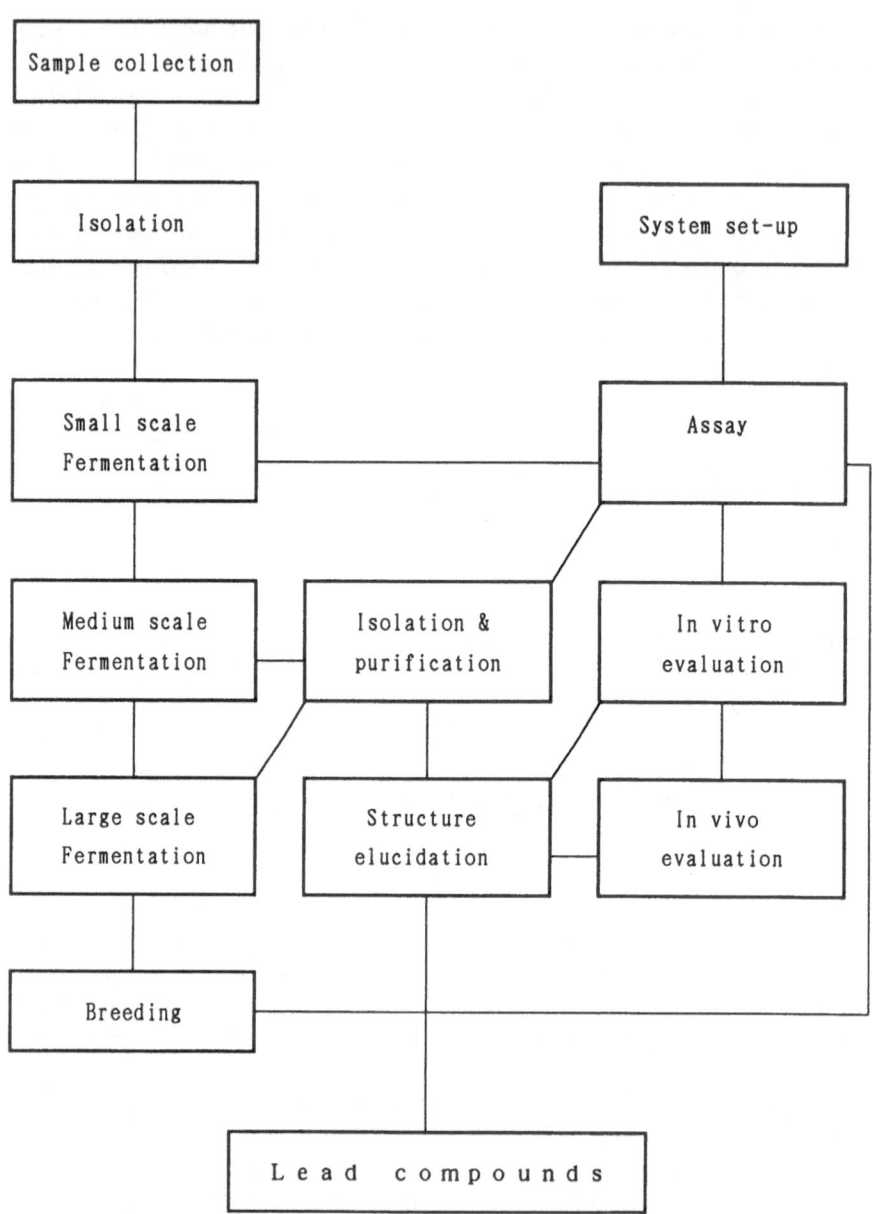

FIGURE 6-6.   Microbial screening flowchart.

The routine assay process is the first to be automated, and it includes pipetting of assay samples; mixing of the reaction mixture; the reaction process; measurement of activity with either liquid scintillation counter, immunoassay reader, or the other monitoring equipment; and data processing. Many useful instruments are now commercially available for these processes. Various manufacturers provide automated systems for pipetting, diluting of samples, and mixing of the reaction solution. For activity measurements, liquid scintillation counters and autogamma counters can be applied. Scintillation counters detect the amplified beam that occurs when radiation activates a scintillator. Liquid scintillation counters are now automated and computerized and can be easily operated. The microplate reader technique is suitable for simple determination of cell growth because of its reliability and rapidness; it is especially useful for the immunoassay, which can detect small quantities of antigen or antibody. Radioimmunoassay, enzyme immunoassay, and fluoroimmunoassay are particularly indispensable for current screening assay systems, including those for measuring the inhibition of cell or virus growth, enzyme inhibition, and protein–protein receptor binding. All of the assay equipment, such as simple photometers for microplates, are now highly developed and automated, and automated equipment for these systems are available from various companies, for instance, Beckmann, Biorad, Biotek, Flow-Titerteck, Hitachi-Corona, Labsystem, and Perkin-Elmer. The denoted instruments have also been used in clinical diagnostics, but they detect only a minor portion of specific compounds in a complex mixture of fermentation broths without special pretreatment. Screening methods should be designed for easy equipment adaptation. For data processing of the assay results, equipment can be connected to a personal computer through RS232C. The binary data are easily handled and processed as ASCII data. Therefore, it is not necessary to develop a specific automated system for the screening assay process.

For the chemical analysis, suitable steps for automation are isolation and chemical detection procedures. The most useful chromatography is probably high-performance liquid chromatography (HPLC). The automated system, provided by various companies, consists of an automatic sampler, pump, injector, column, detector, fraction collector, and data processor. Since the system is highly integrated, commercially available units can be easily applied to desired automated objectives. For the physicochemical characterization step, NMR, mass spectroscopy (MS), IR spectroscopy, and a UV photometer can be used.

The classic method for isolating microorganisms from natural sources, such as soil samples, begins with suspending a small amount of soil, followed by serial dilution. The soil suspension is pipetted and spread over the surface of certain selective isolation agar plates. After incubation for several days, colonies grown on the plates are picked and transferred to an agar slant. Visual selection, the most important process for the isolation of strains, requires extensive experience. For fermentation in tubes, bottles, or flasks, a small portion of spores or mycelia is aseptically inoculated into sterilized media. Inoculated media are cultured on a reciprocal or rotary shaker and allowed to grow and produce secondary metabolites. After several days of fermentation, cultured broths are centrifuged or filtered in a process called *harvest* or *broth-out* to separate the supernatant from the mycelia. Fractions of harvesting culture are heated or extracted with organic solvents and subjected to assays. Dr. R. B. Sykes (personal communication) presented results at the Interscience Conference on Antimicrobial Agents and Chemotherapy (ICAAC) in the early 1980s, and stated that he and his staff cultured more than one million strains of bacteria in order to discover a series of monobactams. Automation seems to be a logical choice for research in this area, and includes preparation of culture media, picking isolates, inoculation, cultivation, harvest, and the preparation of assay samples (Fig. 6–7). Automated equipment for medium preparation would require automated autoclaves, agar dispensers, and various other pipetting and dispensing units. For example, the automatic agar dispenser is very useful when more than 100 agar plates have to be prepared, because it automatically unloads the petri dishes, dispenses the melted agar medium, mixes the poured agar, and loads the petri dishes.

There are also various types of shakers and fermentors for cultivation or fermentation. In addition, a specialized rotary shaker on which small vessels are cultured after the microorganisms are automatically inoculated has been developed (unpublished matter).

The picking up of microorganisms is, in fact, the same process as the inoculation of liquid medium for primary screening. For this procedure, microbiologists use toothpicks or inoculation loops. If they inoculate 100 strains, the process has to be repeated 100 times. This tedious process is also essential for breeding experiments, such as single-colony isolation or mutation trials. Fortunately, several instruments are available for the procedure; for example, Olympus, Tomtec, Toyo-Sokki, and Hitachi all provide systems that transfer colonies from agar plates. Also, scientists at the Nippon Roche Research Center, in collaboration with Hitachi Electronics Engineering Co., have decided to develop a system more integrated than Hitachi's system CT3000.

The quantitative preparation of assay samples is another bottleneck in microbial screening. Most assays require only 10–100 μl of culture liquid, so 96-well titration plates or other small tubes are routinely used. However, varied assay systems should be run in parallel to improve the probability of success. If ten assay systems are run simultaneously, ten identical samples

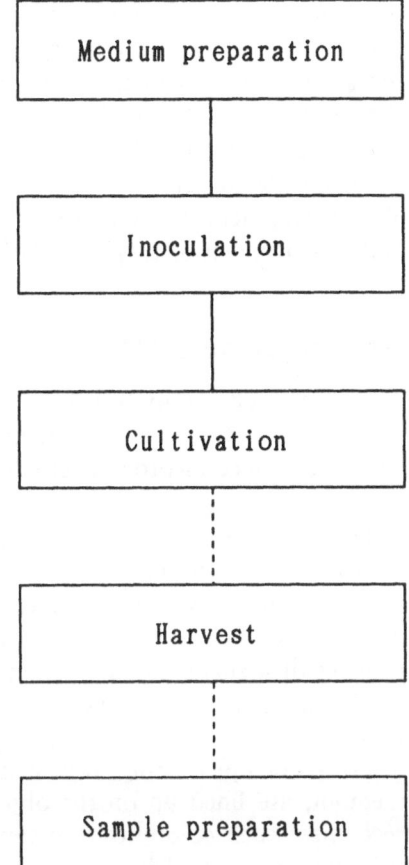

FIGURE 6-7.   Generalized procedure of
cultivation.

have to be prepared for each strain (one broth). If an automated dispenser is used, the same size tubes are required for the unloader and loader. Therefore, a need exists to develop an automated pipetting system to dispense samples from larger tubes to smaller tubes. In addition, the robot should include a cylinder coordinate system that takes up a limited amount of space, since the unit will often be installed in a cold chamber.

AUTOMATIC COLONY TRANSFER SYSTEM

Basic Concept

To screen for a microorganism with a target function, many colonies must be isolated and transferred to various media for determining function and production. The basic work, the morphological analysis and transfer of colonies, has mostly been done manually and requires a substantial amount of labor and time. To cope with this, Hitachi has designed an automated colony transfer and analyzing system (CT2000) that recognizes and analyzes the quantity, size, shape, and color of colonies and subsequently transfers the appropriate colonies. Hitachi also developed a mass production instrument (CT3000) with white and black recognition and mass-transfer functions in order to transfer colonies after mutation and gene manipulation. The development of these instruments has led to quantification, accuracy, and enhancement of aseptic isolation and transfer (Nukumi et al. 1987).

Overview of the CT3000

The automated colony transfer system (CT3000) recognizes the pattern of colonies growing on an agar medium on the source side and automatically transfers each colony to a specified location on the agar media on the object side.

For the manual transfer of colonies, a flame-sterilized Nichrome wire with a 1- to 2-mm-diameter loop, or steam-sterilized disposable wooden stick is customarily used. For automated transfer, a 0.5-mm-diameter stainless steel needle is utilized. This needle is replaced with a new one after the transfer of a single colony. The used needles are collected, washed, sterilized, and then used to replenish the needle stocker while the system is running.

For replica processing, four petri dishes, each containing different agar medium, are lined up on the object side. The ITV monochrome camera recognizes each colony on the source side by its size and coordinates. A colony is transferred by moving the needle holder, which holds a transfer needle within the 3-dimensional space of the apparatus. Various transfer conditions are arranged into readily accessible modes for upgrading operability (Fig. 6–8).

The CT3000 has been tailored for transferring colonies in large quantities, and its operation is simple. In order to automate several steps in our microbial screening, however, modification of CT3000 was needed to achieve the following points:

- The researcher should be able to make adjustments in the robot for any type of microorganism
- Colonies grown on the agar medium need to be selected first and then transferred
- Slant culture cannot be used in an automated system for the isolation of microorganisms
- Liquid media need to be inoculated for primary screening

Source side

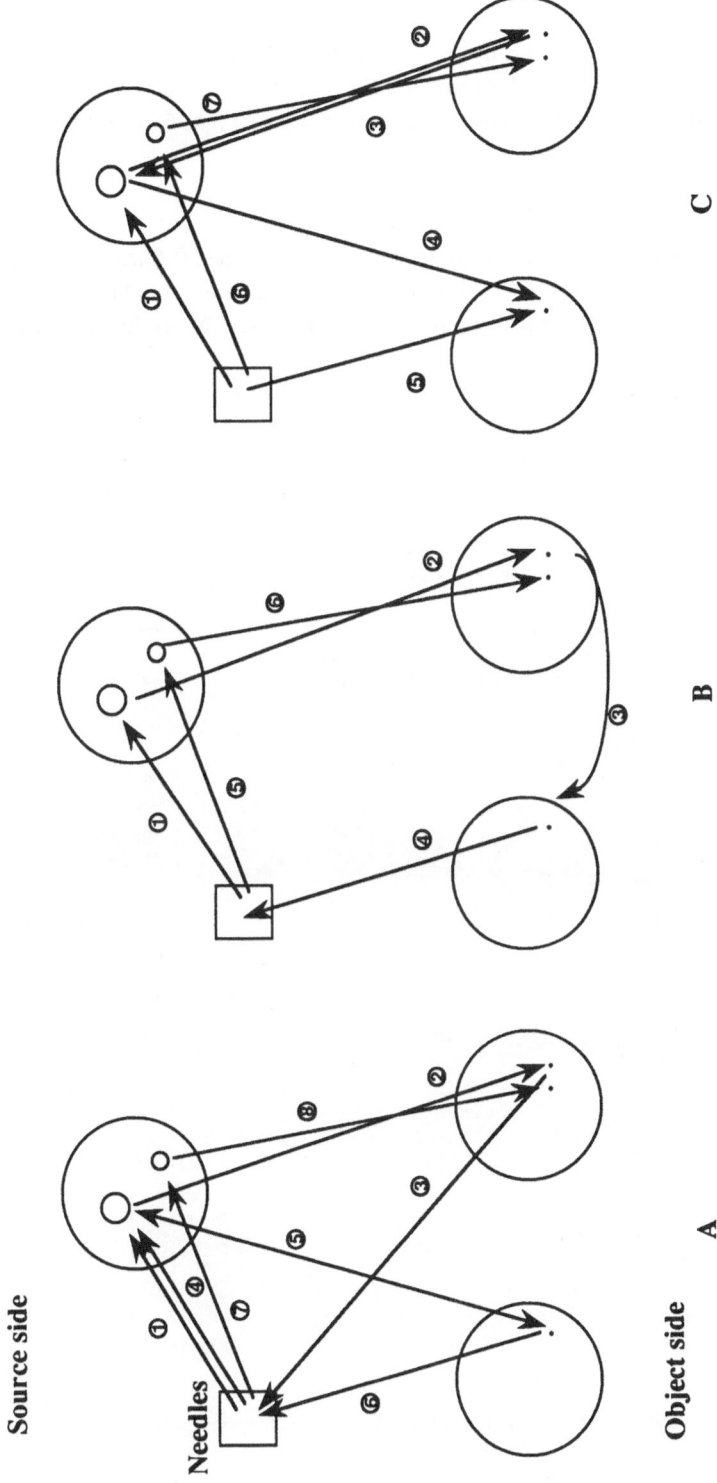

Needles

Object side

A   B   C

FIGURE 6–8. Transfer needle operation and exchange mode. (*A*) Basic operation; (*B*) simple operation; (*C*) reciprocal operation.

The Nippon Roche research group and the engineers of Hitachi concentrated on improving CT3000 and in the process successfully developed a new automated transfer system, the CT4000 (see Figs. 6-9 and 6-10).

Design of Transferring Needles

The CT3000 uses smooth cylindrical stainless needles, 0.5 mm in diameter. At first the effectiveness of the smooth needles was checked. Table 6-1 shows the results of these needles being used for a preliminary manual

FIGURE 6-9.   Automated colony transfer system, CT4000.

TABLE 6-1   Manual Transfer Accuracy with 0.5-mm Smooth Needles

Type of Microorganisms			Growth on/in	
Taxon	Spores	Colony Type	Agar Medium	Liquid Medium
Fungus	−	Hard	+	−
Fungus	+	Wet	+	−
Fungus	+ +	Floccose	−	−
Fungus	+ +		+	+ +
Fungus	−		−	−
Fungus	+ +		+ +	+ +
Fungus	−	Hard	−	−
Fungus	±	Small	−	−
Fungus	+ +	Small	+	+ +
Fungus	+ +		+ +	+ +
Act	+ +		+ + ·	+ +
Act	−		+ +	+ +
Act	−		+	−
Bact		Soft/Wet	+ +	+ +
Bact		Soft/Wet	+ +	+
Bact		Soft/Wet	+ +	+ +

Note: Act = Actinomycetes; Bact = Bacteria; − = none; ± = scarce; + = good; + + = excellent or abundant.

transfer of the colonies into a liquid medium or agar. Each colony was picked with vertical movements, just as the CT3000 system does. All the manually transferred bacteria grew well in liquid media and on agar plates, whereas one strain of actinomycetes did not grow on the agar plate and several fungi did not grow either on agar or in liquid. Colonies of bacteria and yeast are usually wet and sticky; therefore, they can easily be transferred with smooth needles. When dry colonies, such as poorly sporulating fungi or actinomycetes, were isolated, they were picked with a flame-sterilized, bent, or L-shaped needle, or a sterile wooden stick and then transferred to the agar media. Since only a small portion of the colonies is scraped off the agar plates by hand, the requisite movement is nearly impossible for a robot to duplicate. These experiments clearly indicate that it is necessary to improve the shape of the needles or their movements.

To solve this problem, Hitachi engineers designed and constructed five types of needles, which are shown in Figure 6-11. Type 2 is a smooth needle with a 1.0-mm diameter, type 3 is a hollow tube, type 4 has a gap, type 5 has two holes bored through the length of the needle, and type 6 is grooved. The needles were designed to retain mycelia. Poorly sporulating actinomycetes and fungi were used for testing the needles. Although needles 2 through 5 effectively transferred colonies of actinomycetes, only the type 6

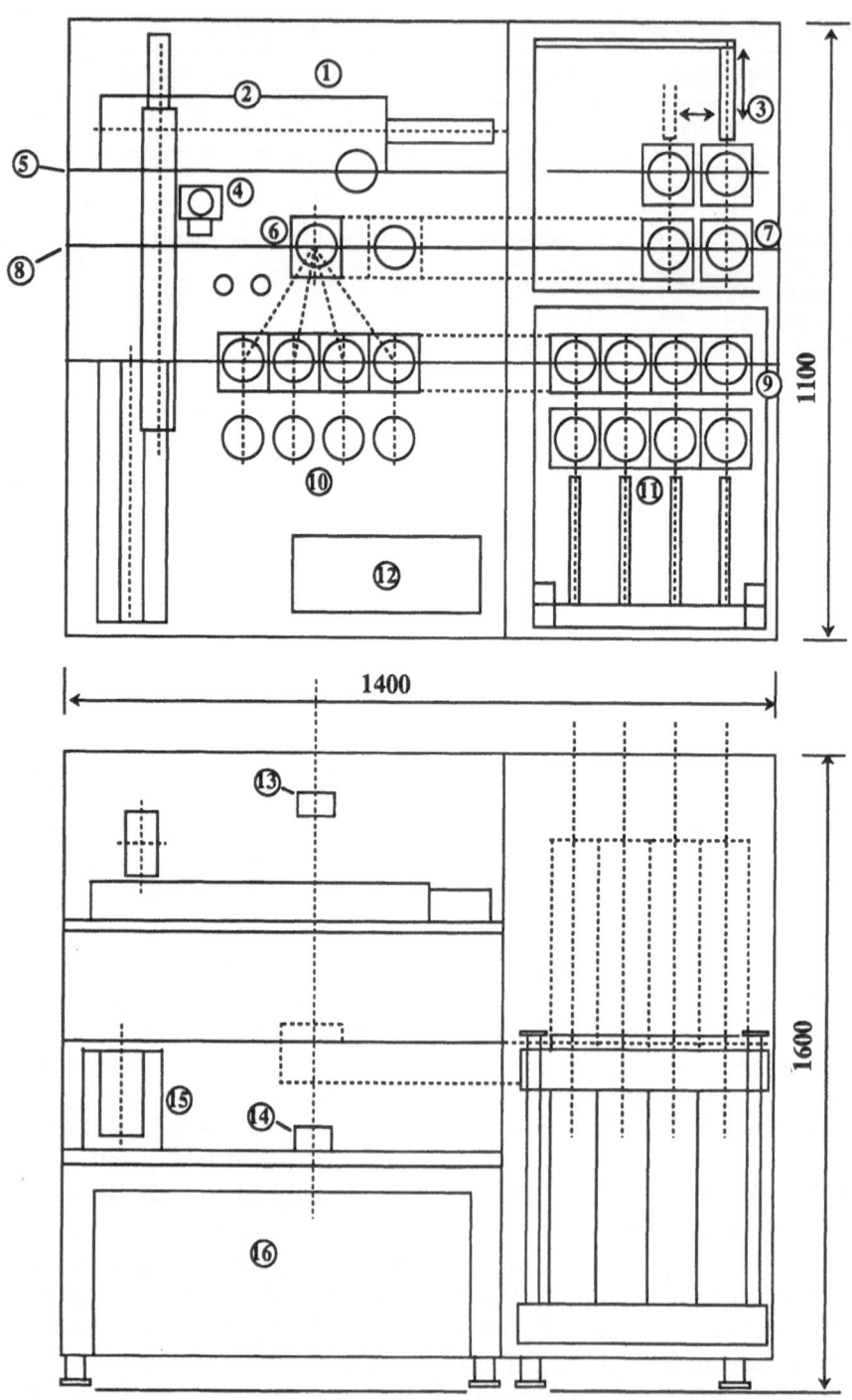

FIGURE 6-10 (at left).   Configuration of the CT4000. (1) Transferring section; (2) X-Y-Z table; (3) source loader section; (4) needle chuch section; (5) source dish transport line; (6) cover removing section; (7) source dish transport section; (8) object dish transport line; (9) object dish transport section; (10) object dish cover removing section; (11) object loader section; (12) operation panel; (13) upper ITV camera; (14) lower ITV camera; (15) needle supply/storage section; (16) control box.

with grooves was sufficient for transferring fungal colonies (75% of poorly sporulating fungal colonies were transferred, as shown in Table 6–2). Thus, the process succeeded in improving the accuracy of colony transfer.

Improvement in Accuracy of Inoculation

In addition to a change in the design of inoculation needles, further improvements were necessary to increase the accuracy of inoculation. When CT3000 picks up microorganisms and inoculates media, the arm of the right-angle coordinate robot moves up and down along the Z-axis, which results in the formation of pinpoint colonies. When microorganisms are inoculated onto the agar slant or into the liquid medium by hand, a loopful of culture is usually streaked lightly back and forth on the surface of the agar or in the liquid. If back-and-forth movements are added to the robot,

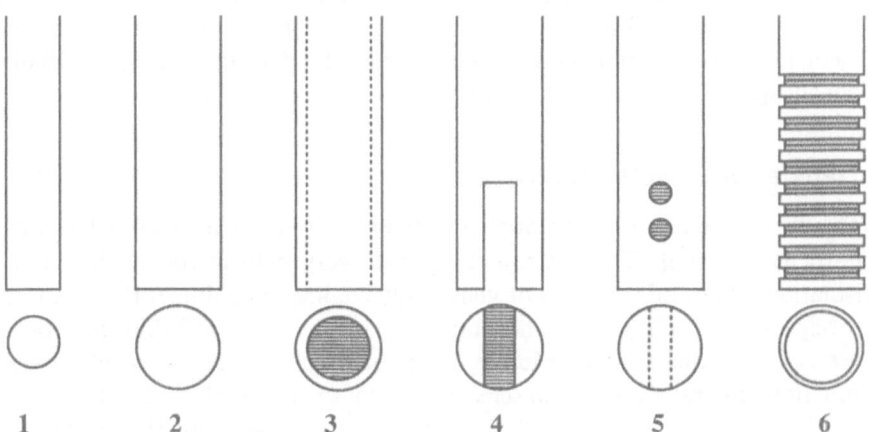

FIGURE 6–11.   Various types of transfer needles. (1) Original 5-mm diam. needle; (2) smooth needle, 1.0 mm diam.; (3) hollow tube; (4) needle with a gap; (5) needle with two holes; (6) grooved needle.

TABLE 6-2   Effect of Needle Design on the Accuracy
of Transferring Poorly Sporulating Fungi

Needle	Medium	Transferring Accuracy Rate (%)
1	Agar	20
	Liquid	40
2	Agar	50
	Liquid	30
3	Agar	20
	Liquid	10
4	Agar	10
	Liquid	0
5	Agar	40
	Liquid	30
6	Agar	90
	Liquid	50

the inoculation accuracy would most likely be increased, particularly in the case of liquid medium. Therefore, an additional function for inoculation called a *scrub function* (back-and-forth movements) was developed. In addition to the simple movement along the vertical axis, this function enables horizontal (*X*-axis) movement of the inoculation needle after the needle touches the agar surface and pierces into it. Hopefully, this function would allow for movements similar to those done manually. This function can be repeated one to 99 times. When the same colony has to be picked up a second time, the needle moves back and forth perpendicular to the previous movement, to avoid touching the same area of the colony. This improvement provided for inoculation of any type of microorganisms, especially poorly sporulating fungi, more precisely into solid or liquid media.

Transfer of Selected Colonies

The prototype instrument transfers every colony on an agar plate [see Fig. 6-16(A) and (C)]. This instrument will be very helpful for single-colony isolation trails and selection of genetically engineered colonies that contain foreign DNAs. However, when microorganisms from soil-dilution plates are isolated, the desired colonies are selected and picked. An additional function for transferring the selected colonies was therefore developed.

The TV camera positioned above the source stage scans the surface of petri dishes in CT3000 to recognize every colony. The image of colonies on the agar surface is interfaced with the microcomputer as binary data. If the position of preferable colonies is marked on the reverse side of petri dishes with an appropriate marker pen, another TV camera, installed below the source stage, can be programmed to recognize these markers. Hitachi's engineers found that white markers were the best.

Transfer to a Six-well Plate

For isolation of microorganisms, the agar slant is customarily used; however, since automation with the slant was impossible, a six-well titration plate, in which each well (resembling a small petri dish) contained 2 to 3 ml of sterile agar medium (Fig. 6–12), was used. The needle picks up a selected colony from the source petri dish and transfers the microorganism to three points in each well of the plate. After the inoculation, the six-well plates are incubated for an appropriate time to allow for growth. Since the colonies grown in several wells of the plates might be contaminated, or growth may be insufficient, selection of a desired well is necessary. Thus, a sorting function was added for the selection of desired colonies. The process was accomplished by pasting a piece of aluminum tape on the reverse side of each unfavorable well, thus signaling the robot to disregard the marked wells and to recognize only the unmarked wells as containing colonies to be transferred to a new six-well plate. The final plate is used as a master plate that is similar to a conventional agar slant culture (Fig. 6–13).

Automated Inoculation of Liquid Media

For microbial screening, cultures in liquid media are often used for checking specific activity in broth. The process requires a robot to inoculate a

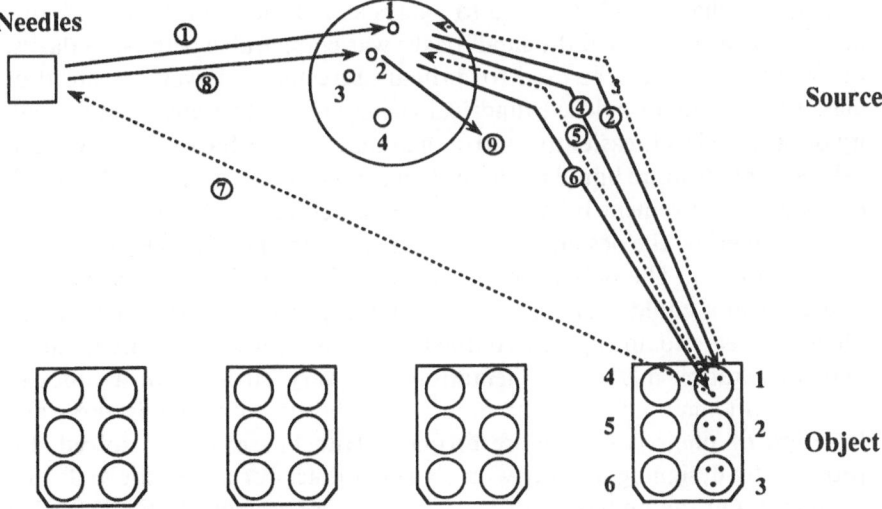

FIGURE 6-12.   Transferral to six-well microplates.

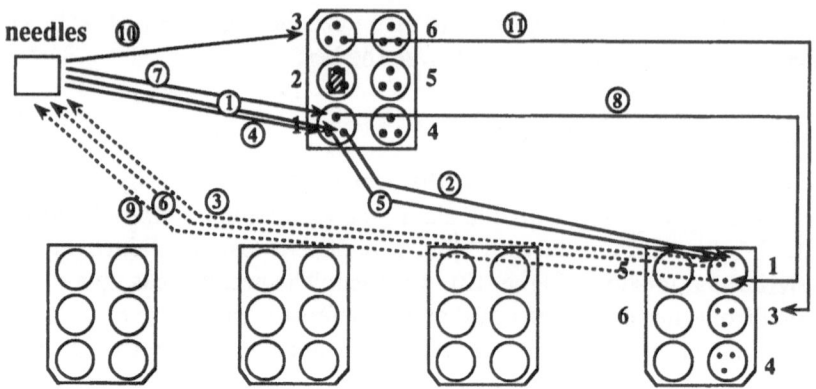

FIGURE 6-13.   Sorting function of the transferral to microplates.

liquid medium on a small scale. However, the motions needed for the inoculation of a liquid medium are different from those for a solid.

Initially, a culture bottle had to be designed for primary screening. For small-scale fermentation, test tube cultures or small (50- to 100-ml size) Erlenmeyer flask cultures can usually be employed, although flask cultivation is easier for scaleup fermentation. The six-well plate has six wells, each approximately 4 cm in diameter. If six small flasks are set in a specialized container patterned to resemble six wells in the plate, the inoculation process could also be done by the robot. The fermentation efficacy of a 50-ml flask (4 cm diameter) was checked. Several known antibiotic producers were introduced into the flasks (each containing 10 ml of liquid medium), which were shaken at 27°C for 3 to 5 days at 160 rpm. Since the antibiotic activity of the cultures in the small flasks was comparable with large flasks, the small flasks could be used for primary screening. However, instead of glass flasks, polypropylene cylindrical vials were used because (1) the quality of plastic vials is easier to control than that of glass flasks, (2) a cylindrical vessel holds more liquid medium than a conical one, and (3) polypropylene is more convenient because it can be autoclaved or disposed.

Subsequently, the design of the container for the plastic vials (Fig. 6-14) was considered. The design of the previous six-well plates was utilized so that the robot could perform the inoculation procedure. Although cotton plugs or plastic foam caps are routinely used for preventing contamination and for aeration in flask fermentation, it is very difficult for a robot to remove soft material from the mouth of a bottle without mistakes, and insertion of plugs is even more difficult. Hitachi's engineers solved this problem by designing a cover with six small holes for six plastic vials in a container. On the underside of the cover, six large, soft-plastic disks that are sufficiently large to cover the hole and large enough to cover the mouth

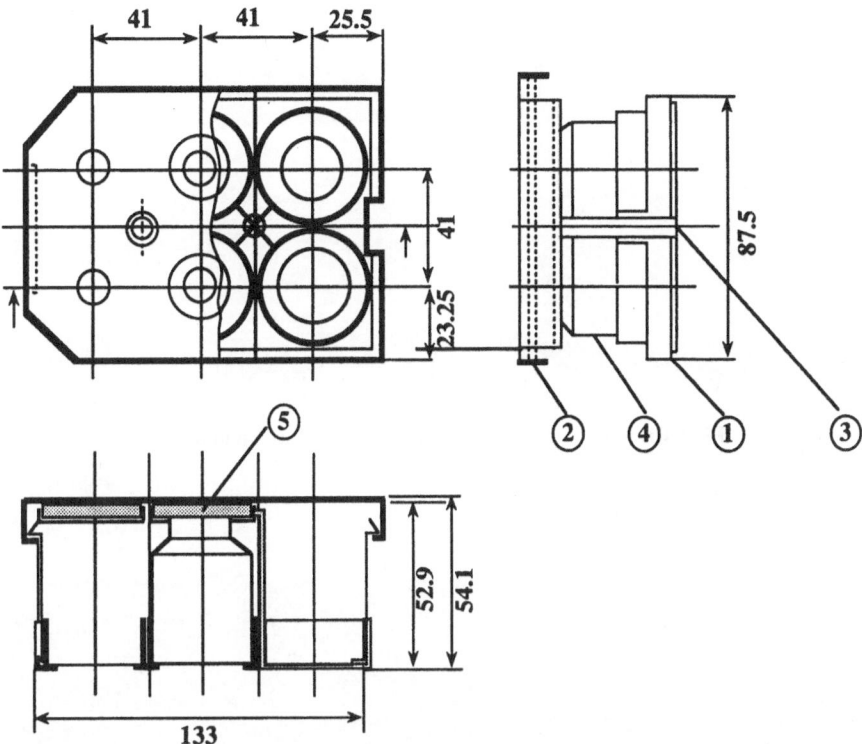

FIGURE 6-14.   Vessel container assembly. (1) Vial case (bottom); (2) vial cover (top); (3) leaf spring; (4) vial; (5) plastic foam (cover).

of the vial are secured. The upper part of each cover is attached to a lower container by two stainless leaf springs. The two arms of the robot push the springs until the top pops up slightly so that the cap board can easily be pulled and opened with a vacuum for the transfer of colonies into liquid media (Fig. 6-15). After the inoculation, the cover is pushed down and locked with the stainless springs. The whole container with six plastic vials is shaken by a specialized rotary shaker. A function to drop the inoculation needle into the liquid medium has been added in order to increase the inoculum size (Ohkawa et al. 1989).

Evaluation of the System

After a little over a year of development, Hitachi completed the new automatic colony transfer system, CT4000 (see Figs. 6-9 and 6-10). The accuracy of transferring various types of microorganisms with this system was examined (Figs. 6-16 and 6-17). Because of various improvements, the re-

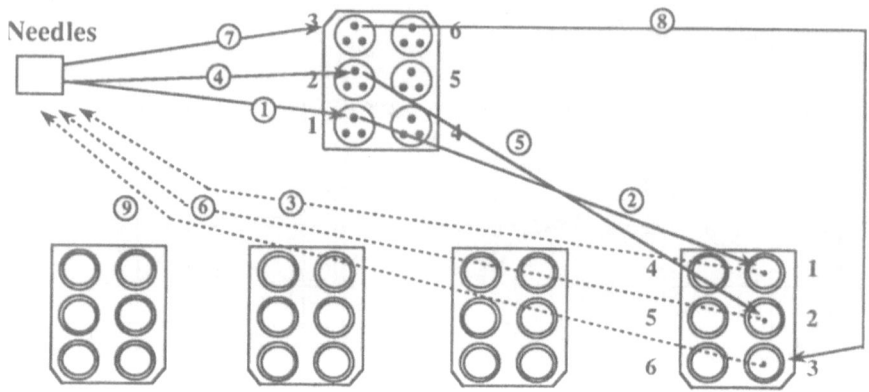

FIGURE 6-15.   Sequence of the transferral to liquid media.

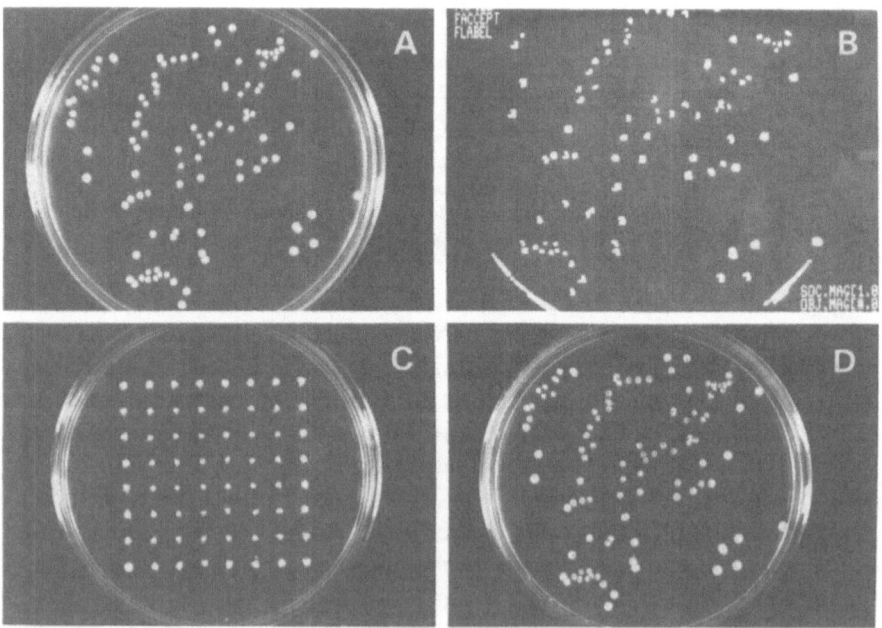

FIGURE 6-16.   Precise inoculation of *Kluyveromyces fragillis* ATCC8635. (*A*) Source dish; (*B*) binary data on CRT; (*C*) object dish after incubation; (*D*) source dish after the transferral. (See the center of each colony.)

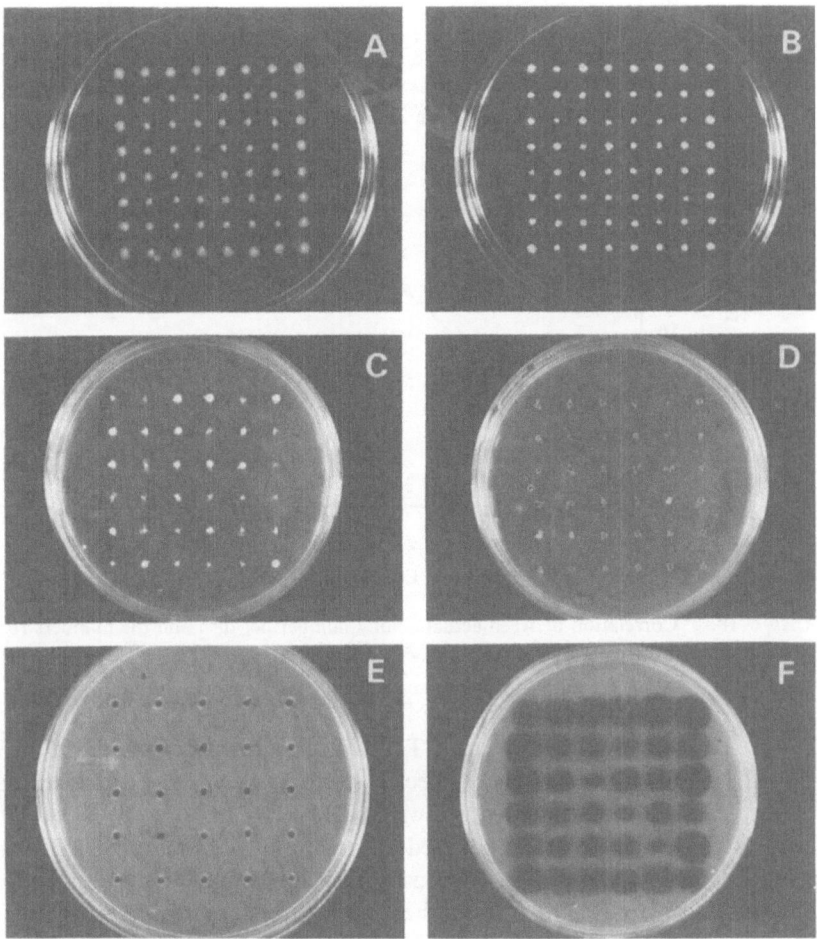

FIGURE 6-17.   Precise inoculation of various taxa. (*A*) *Excherichia coli* NIHJ; (*B*) *Staphylococcus aureus* 209P; (*C*) *Streptomyces griseus* ISP5236; (*D*) *Micromonospora chalcea* NR0029; (*E*) *Cladosporium spherospermum* NR6306; (*F*) *Phialophora verrucosa* MTU23003.

sult showed homogeneous accuracy that was not dependent on the types of colonies transferred; that is, not only soft colonies but also hard or nonsporulating colonies were precisely transferred and grown on new media.

The CT4000 system precisely recognized and transferred up to 200 colonies grown on one petri dish (Fig. 6-18). This system will enhance our ability to isolate a large number of colonies in less time than it took with the

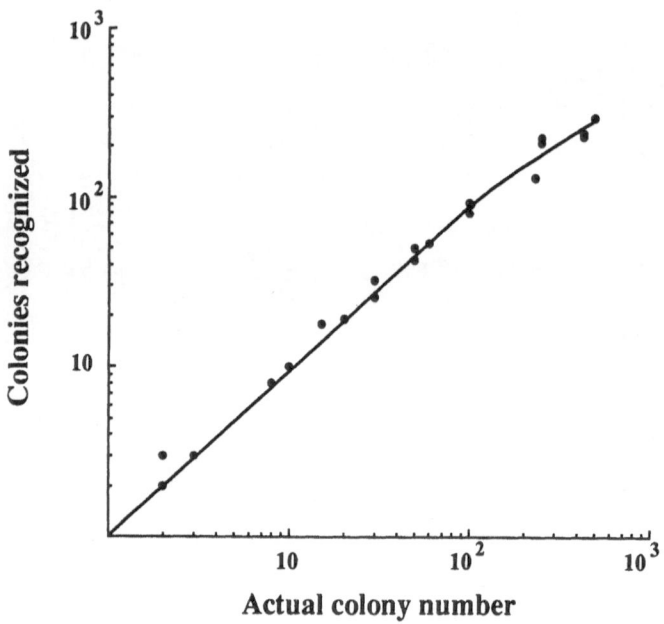

FIGURE 6-18.   Correlation between actual colony number per dish and the numbers recognized per dish when *Kluyveromyces fragilis* ATCC8635 was used.

CT3000. All the functions of the CT4000 are summarized in Table 6-3, from which researchers can easily select various combinations of functions. The speed and processability are shown in Table 6-4 and Figure 6-19. The actual transfer operation is conducted according to a combination of various modes. Transfer speed and processability are largely dependent on the needle exchange mode and the number of transfers per dish. However, if 40 source petri dishes were used with 50 colonies per dish (total 2000 colonies), it would take approximately 4 hours to transfer all of the colonies by the basic mode. The CT4000 is therefore one of the most useful tools for biotechnology researchers who want to select colonies containing foreign DNA or want to isolate auxotrophic mutants from a large number of clones.

DISPENSING ROBOT

The assay sample preparation requires complex movements, and full automation may not be economically feasible. Startup costs are the primary deterrent to management's decision to utilize laboratory automation. A careful study of the step-by-step procedure in Figure 6-20 supports the con-

TABLE 6-3   Function List of CT4000

Source and object dishes	Petri dish to petri dish
	Petri dish to six-well microplate
	Six-well microplate to microplate
	Six-well microplate to bottle
Recognition mode	Colony recognition with upper camera
	Marker recognition with lower camera
	Recognition of aluminum tape on the reverse side of microplate
Transfer mode	To one medium
	To 2 kinds of medium
	To 3 kinds of medium
	To 4 kinds of medium
Transferring method	1 point/dish
	2 points/dish
	3 points/dish
	16 points (4×4)/dish
	25 points (5×5)/dish
	36 points (6×6)/dish
	49 points (7×7)/dish
	64 points (8×8)/dish
	100 points (10×10)/dish
Needle exchange mode[a]	Basic
	Simple
	Reciprocal
Needle movement	Movement along Z-axis
	Scrub along X- and Y-axes
	Needle drop mode (bottle)

[a]See Figure 6-8.

TABLE 6-4   Transferring Speed

Needle Operation and Exchange Mode (see Fig. 6-8)	Object Dish Set Mode		
	4 Kinds of Media (sec/point)	3 Kinds of Media (sec/point)	2 Kinds of Media (sec/point)
Basic operation	6.4	6.5	6.4
Simple operation	2.5	3.0	3.8
Reciprocal operation	3.8	4.1	4.7

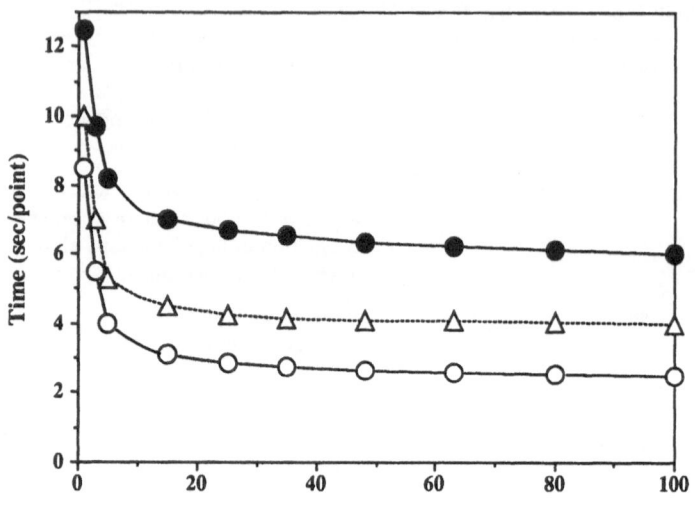

Number of transfer (No. of points/object dish)

FIGURE 6-19.   Processability. Transfer needle operation and exchange mode. —●—
= Basic operation; ----△---- = Simple operation; —○— = reciprocal operation.

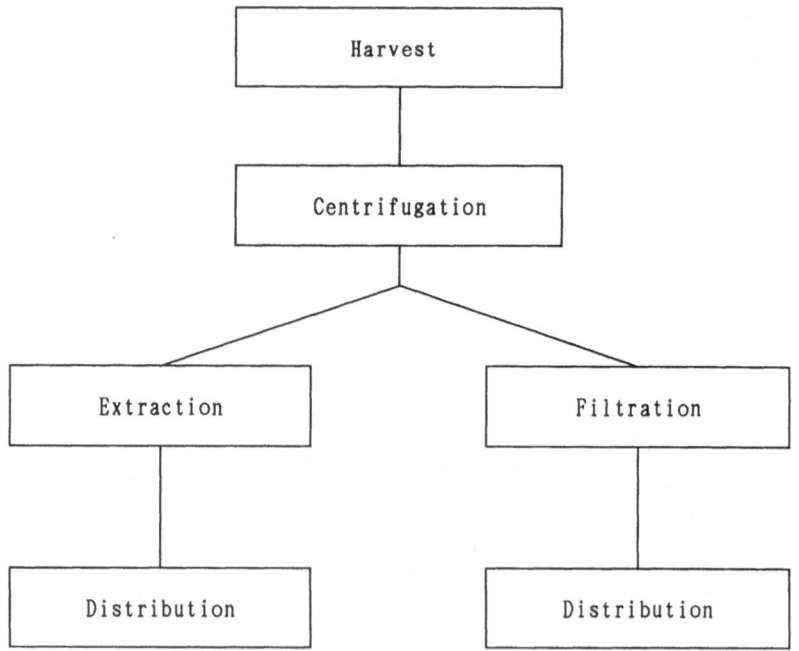

FIGURE 6-20.   Schematic for full sample preparation.

clusion that partial automation systems for dispensing samples should be developed so that researchers can be freed from performing repetitive tasks to spend time on creative and innovative work. Four conditions should be met for the economic feasibility of automation:

- Maintain the existing sequence in the manual procedure
- Use the same-size tubes
- Reduce and simplify the total number of steps in the procedure, that is, the size, number, or type of racks may need to be changed
- The space required should be no greater than one cubic meter to enable the use of a small cold chamber

Equipment from Labomaster Jr. designed by Umetani Precision Co. was chosen because it met the denoted conditions (Fig. 6–21). No comparable equipment was commercially available that dispensed a large number of liquid samples from larger tubes to smaller tubes.

Equipment Setup

For future expansion of the automation setup, such as the addition of a filtering unit, the same type of stand was designed to hold cassette-type racks. Two-tiered, slightly straddled stainless steel stands are positioned

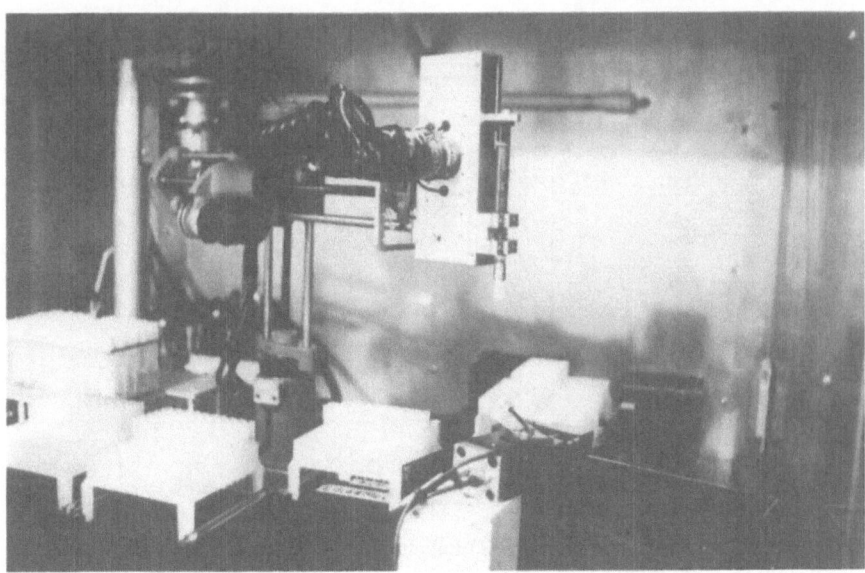

FIGURE 6-21.   Dispensing robot, Labomaster Jr.

along the diameter of circles concentric to the robot in a 100-cm³ cold chamber. The racks were set on nipples so that they would be securely anchored. To reduce manual work time, disposable tubes and tips were used. The layout of the distribution tier (Fig. 6–22) consists of sample racks (4A and 4B), distribution racks (5A–5E), a tip rack (3), a tip remover (7), and tip sensor (8). A distribution menu was used to make functional assignments from the CRT during setup. One of the parameters involved setting the plunger position (0 to 1000 ml) for absorption during the distribution process. The sample rack on a stand is held in place with four pins, so that it remains in the same position during the rack exchange. The distribution rack, made of polypropylene by injection molding, is a cassette type designed so that the samples in the tubes can be processed without removing the tubes from the rack.

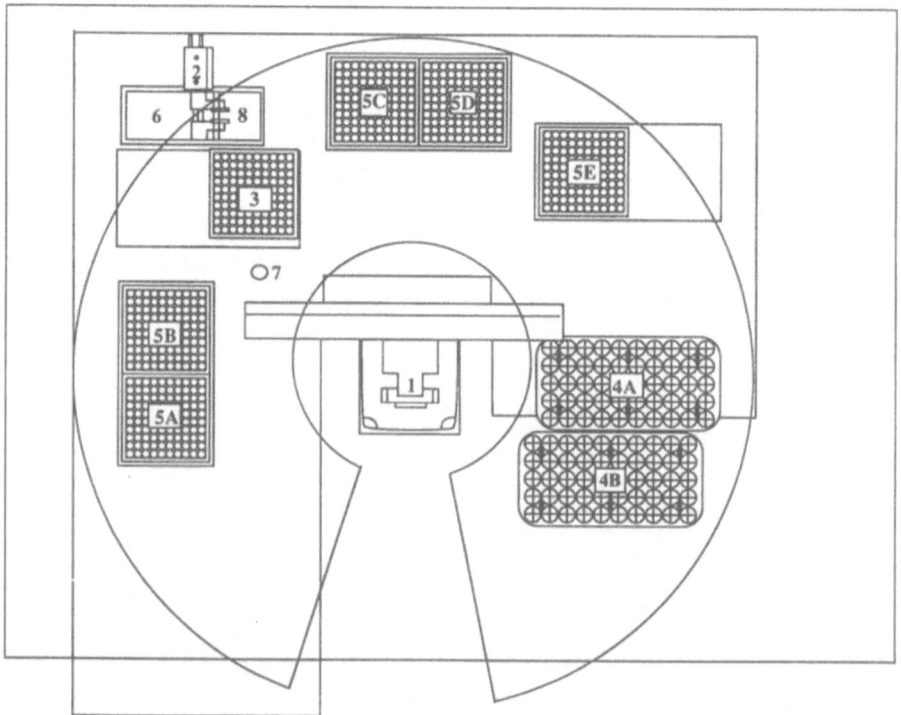

FIGURE 6-22.   Distribution system. (1) Labomaster Jr.; (2) mechanical clamp to remove tips; (3) tip rack; (4) sample rack; (5) distribution rack; (6) container for used tips; (7) tip remover (8) tip sensor.

Distribution System

The objective of the distribution system is to remove sample liquid from 100 larger tubes and transfer them into 100 smaller tubes without contaminating the sample. The smaller tubes were set in 94- × 94-mm racks (Fig. 6–23). The robot will pick up the sample liquid using a syringe. The amount of sample in the syringe is calculated by the CPU, and one of the following actions is taken: (1) if the volume in the syringe is greater than or equal to the specified distribution volume, then the specified volume is distributed, and (2) if the volume in the syringe is not sufficient for the next distribution, then more sample needs to be pipetted and the specified volume distributed. For example, each tube in the first rack requires 400 μl, and the second requires 500 μl. If the syringe contains 800 μl and it distributes 400 μl into a tube in the first rack, only 400 μl is left, so the syringe must return to the sample rack to pipette the required volume of liquid. Contamination caused by the sample dripping while being transported and volume errors can be prevented by first pipetting the sample and subsequently sucking up a measured amount of air before the robot arm is moved. Pipetting accuracy was attained as shown in Table 6–5. The total quantity of the sample liquid in

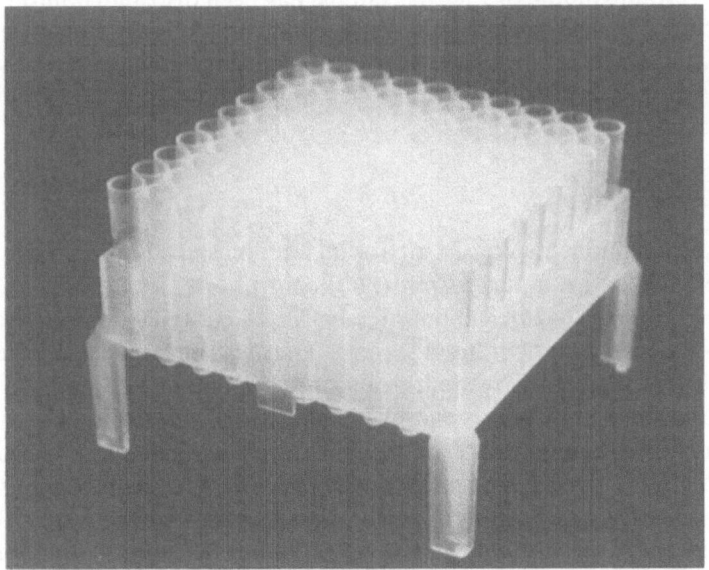

FIGURE 6-23.   Distribution rack with tubes.

TABLE 6–5   Pipetting Accuracy and
Reproducibility with 50 Percent Methanol

	Pipetting Volumes	
	---	---
Trial	1.0 ml (g)	0.5 ml (g)
1	0.948	0.442
2	0.949	0.459
3	0.950	0.461
4	0.954	0.461
5	0.955	0.460
6	0.950	0.450
7	0.950	0.462
8	0.949	0.457
9	0.953	0.459
10	0.954	0.461
Average	0.951	0.457

each rack is shown on the CRT and with input on a 10-key board. The tip is removed by a mechanical clamp after distributing each liquid sample. After a certain volume of the first sample has been distributed into the first smaller tube of each object rack, the second sample is distributed into the second tube of the object rack. The entire distribution process of distributing the samples into 500 tubes takes less than 4 hours.

Main Drive System

The main components consist of robot Labomaster Jr., a personal computer NEC9801, an input/output (I/O) control box, a tip remover, and a teaching box (Fig. 6–24). Labomaster Jr. is a 45-cm-tall-cylinder robot with a DC servodrive that operates a syringe. The arm operates within the cylinder with five degrees of freedom (rotation, horizontal reach, vertical reach, wrist, and fingers). The servoamplifier consists of the main CPU and five auxiliary CPUs that control arm movements by a master–slave relationship. The main CPU selects and sends movement orders to each auxiliary CPU. Slave CPU drives of each servomotor send feedback of the results back to the master CPU. Positioning accuracy is within 0.1 mm. Communication between the CPU and control box controls the tip remover. Feedback from the sensor determines whether action should be taken by the error-control program (see Fig. 6–27).

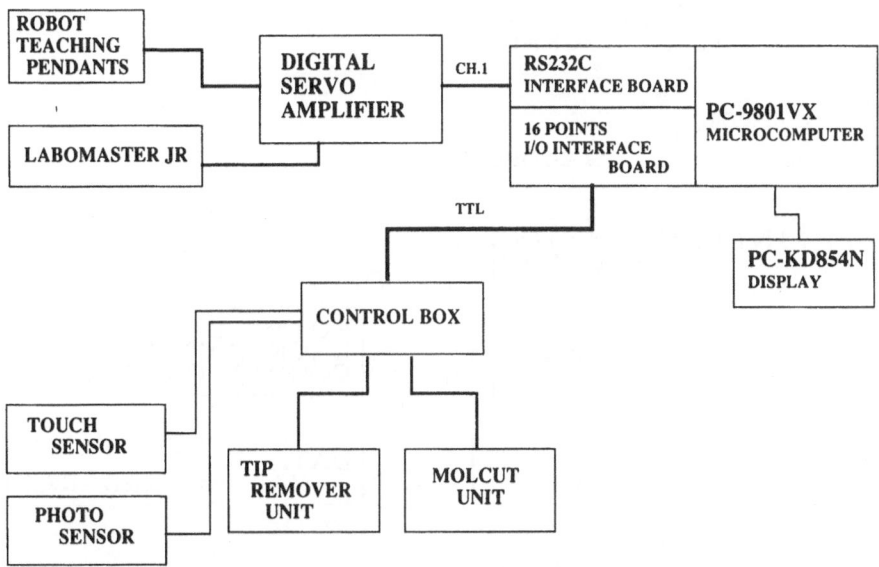

FIGURE 6-24.   Electrical schematic showing setup for distribution system.

Control System Program

The control system program (Fig. 6-25) is written in BASIC and assembly language on MS-DOS. Communication between the CPU and servoamplifier uses the data terminal ready–data set ready (DTR–DSR) handshake method. The communication software was first written in BASIC, but because of frequent errors, it was rewritten in assembly language. The program is designed to enable the user to change the parameters that determine the robot arm placement to compensate for different sizes, number of racks, and so forth. The setup sequence for robot movement is as follows: contact points are registered into the teaching box and transferred/saved in a data disk; the distribution menu on the master CPU is used to input the required conditions, which determine the sequence of movements from point to point and enable the running of the distribution program; and finally, orders are sent to the servoamplifier by a digital control signal.

The positioning control program determines the movement from one point to another. This program converts absolute coordinates into relative coordinates. Thus, two data sets are created: the registered data entered by the operator is converted into point and rack data sets; and the rack data set is the value calculated by the palletizing program for the positions of 100 or 50 tubes based on the specified point within the four corners of the rack. The variance from center position (mouth of tube) in each pipetting

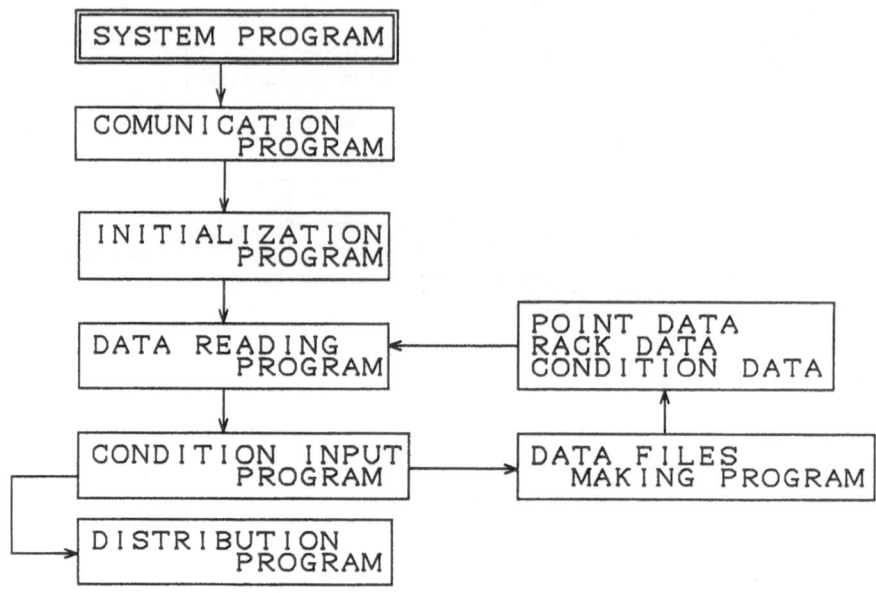

FIGURE 6-25. Architecture of the control program for the system.

task is 0.1 mm. The flow chart (Fig. 6–26) shows the sequence of steps for the robotic movements.

Evaluation and Possible Improvement of the System

During initial development, robotic abnormalities were caused by electrical noise, defective tips, and an unclear sensor. All problems occurred during distribution system operation when the robotic arm was not in its precise position. To handle this abnormality, the following procedure was added to the distribution program to detect and correct or stop an operation by using a sensor on the mechanical clump to remove tips (Fig. 6–27):

1. Syringe hand extends to remover to check robot movement
2. Sensor detects whether the syringe hand is "on" or "off;" Flag 1 or Flag 0 status in the figure means that the tip is either "on" or "off," respectively
3. Clamp removes tip and hand lifts up
4. Sensor detects the presence of a tip in the clamp
5. The clamp opens, releasing the disposable tip, and the tip drops into a container

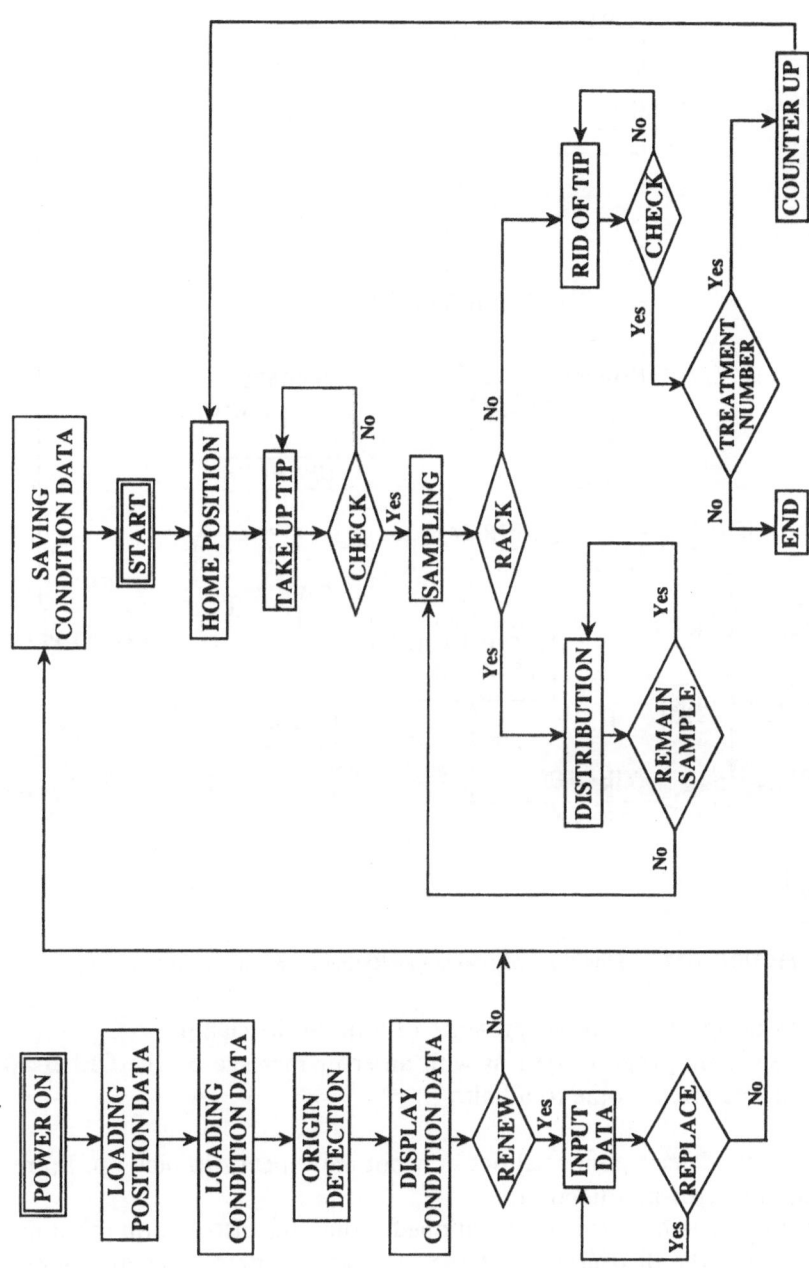

FIGURE 6-26. Flowchart of the distribution system program.

145

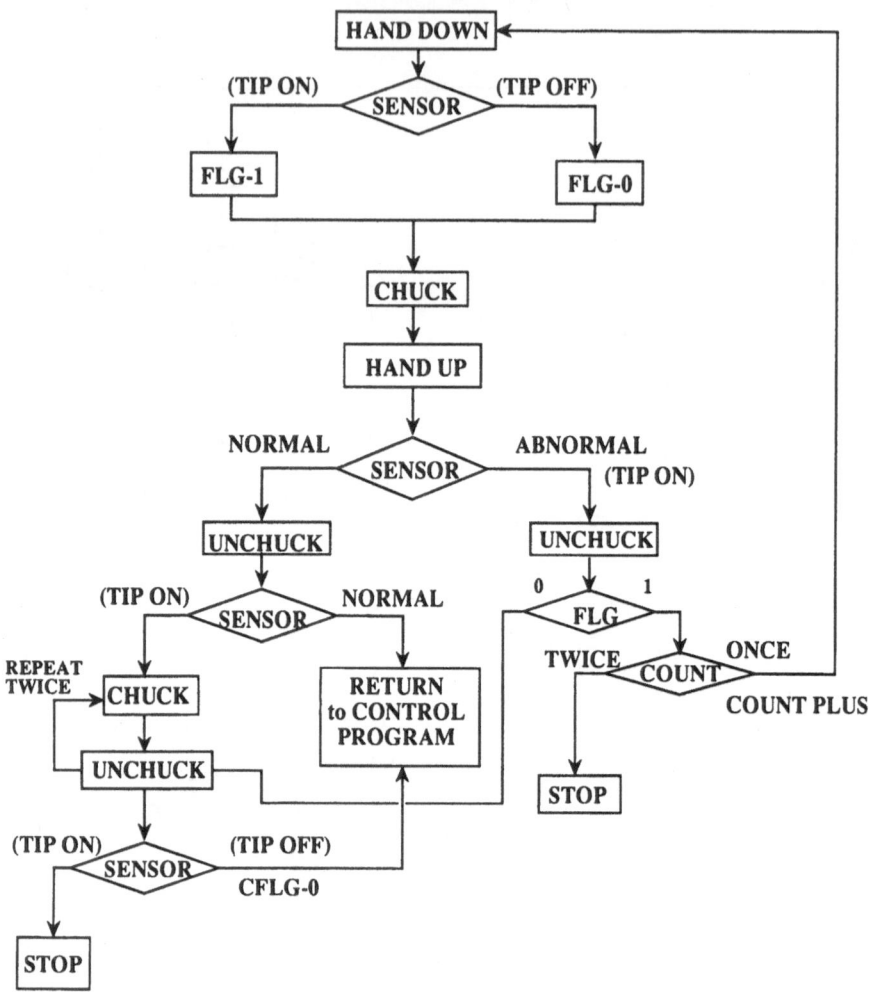

FIGURE 6-27.   Flowchart of system monitoring program for tip detection.

6. Sensor again detects the presence of a tip in the clamp
7. The system program returns with an error message on the CFLG$=0$ and operations come to a halt

This process detects abnormalities in robot movement, tip pick up, tip removal, and error malfunction.

Through the improvements mentioned earlier, the distribution robot has provided for the dispensation of the required amounts of assay samples from 100 larger tubes into 500 smaller tubes with unmanned operation. The robot has cassette-type racks, which are multifunctional.

There are several choices for obtaining assay samples without undesirable high molecular weight proteins. The samples could be extracted with solvent, heated, or filtered through a membrane with a molecular weight cutoff. For a further improvement of the sample preparation process, Labomaster Jr. (Fig. 6-28) will be used for automating the filtration step. A specially designed filtering unit (Molcut Filtering Unit, Millipore Company) is used for filtration. This process is complex and time-consuming compared to sample distribution, since filtering may take twice as long as distribution. The Molcut procedure with Labomaster Jr. is as follows:

1. The syringe picks up the tip and distributes the sample into the Molcut filtering unit, which is connected to a tube
2. The Molcut unit with tube is transferred to the capping unit
3. The cap is picked up from the cap rack and gently positioned onto the Molcut unit

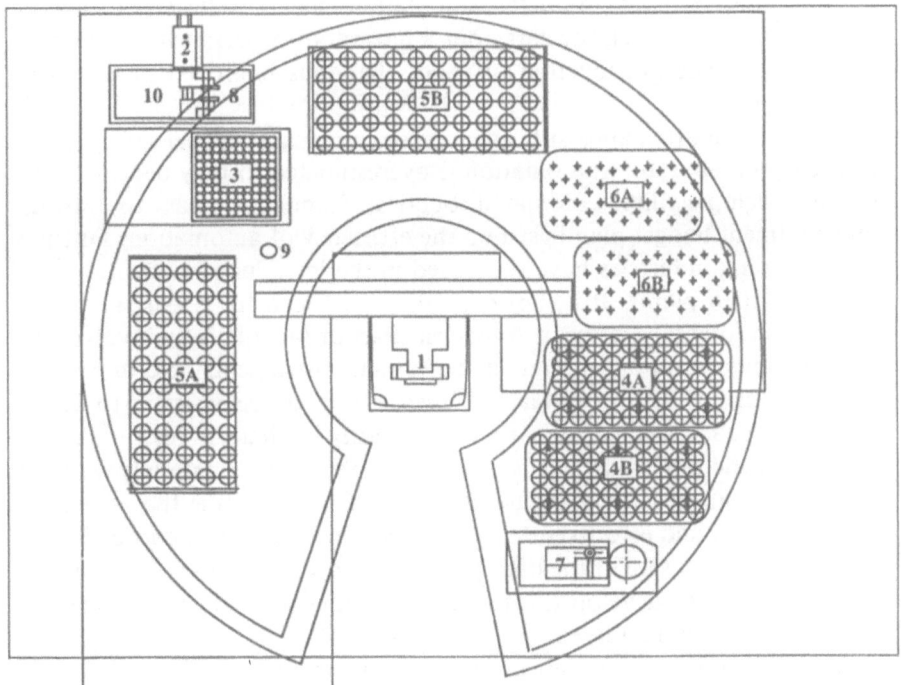

FIGURE 6-28.   Filtration system. (1) Labomaster Jr.; (2) mechanical clamp to remove tips; (3) tip rack; (4) sample rack; (5) filter rack; (6) filtration assembly cap rack; (7) capping unit; (8) tip sensor; (9) tip remover; (10) container for used tips.

4. The rotary head pushes the cap down, and then screws on to tighten the cap
5. The solenoid valve is opened and air is purged through the filter of the Molcut unit
6. The robot picks up the Molcut unit with the tube and returns it to the cap rack

The filtering process is carried out in the Molcut tube while it is in the cap rack. The processing of 15 samples takes approximately one hour to complete. It would be best to complete the sample distribution near the end of the day so that the filtering could be done overnight.

GENERAL COMMENTS ON DEVELOPMENT OF THE AUTOMATED SYSTEM

Large-scale automation is now used for the manufacture of automobiles, plastics, and electrical products. Automation enables industries to manufacture round the clock, or carry out a hazardous process without human intervention. The introduction of robots has thus improved productivity and quality.

Automation often aims at both increased efficiency and effectiveness. If efficiency in laboratory automation is overestimated, it may become difficult to develop credible automated robotics, although it is easy to draw up an investment budget plan based on the efficiency of automation. Furthermore, leading industries have succeeded in the introduction of total automation in their plants. It is, however, rather difficult for research areas to accomplish total laboratory automation. Part of the difficulty involves the lack of standardization of many materials and instruments, and in predicting how long any project will last. Therefore it might be desirable to design an open-ended, feasible system that could easily be changed for future programs. On the other hand, it is difficult to standardize materials within a flexible system, because standardization itself might define the system. If both efficiency and effectiveness could be achieved, the total cost of investment would increase. For the time being, therefore, module-type automated robotics should be developed for several components of the total system in the area of microbial screening or breeding.

When a company considers automated equipment for the first time, it is best to look at commercially available equipment. If a relevant system is not available, an automated system could be developed in collaboration with a manufacturer. Based on past experiences, selecting a cooperative manufacturer that employs highly capable engineers is very important. The manufacturer need not be familiar with the pertinent field. Professor Sakuzo Fukui once stated at a meeting of the Bioindustry Development Center (BIDEC) that if young researchers have an interest in automation, and if they carefully analyze conventional methods, he will be pleased to let them do it.

Finally, laboratory automation should not be considered to save human resources, but for liberating researchers from repetitive and time-consuming experimental work. It is important to emphasize the effectiveness of automated systems, because it will allow scientists to accomplish their innovative research and achieve efficiency.

Acknowledgment

We would like to thank Mr. N. Sugiura and Mr. M. Nukumi, Hitachi Electronics Engineering Company, for kindly allowing us to use their CT4000 data, and for offering suggestions in the writing of this chapter.

References

Borel, J. F., C. Feurer, C. Magnee, and H. Stahelin. 1977. Effects of the new anti-lymphocytic peptide cyclosporin A in animals. *Immunology* 32:1017–1025.

Dreyfuss, M., E. Harii, H. Hoffman, H. Kobel, W. Pache, and H. Tscherter. 1976. Cyclosporin A and C. *Eur. J. Appl. Microbiol.* 3:125–133.

Endo, A. 1979. Monacolin K, a new hypocholesterolemic agent produced by a *Monascus* species. *J. Antibiotics* 32:852–854.

Endo, A., M. Kuroda, and Y. Tsujita. 1976. ML-236A, ML-236B and ML-236C, new inhibitors of cholesterogenesis produced by *Penicillium citrinum*. *J. Antibiotics* 29:1346–1348.

Fukujoji, S., M. Sugihara, H. Imai, T. Soejima, and Y. Odawara. 1987. Trends of biotechnology equipment. *Hitachi Hyoron* 69:289–294.

Freidinger, R. M. 1989. Cholecystokinin and gastrin antagonists. *Medicinal Res. Rev.* 9:271–290.

Goetz, M. A., M. Lopez, R. L. Monaghan, R. S. L. Chang, V. J. Lotti, and T. B. Chen. 1985. Asperlicin, a novel nonpeptidal cholecystokinin antagonist from *Aspergillus alliaceus*. *J. Antibiotics* 38:1633–1637.

Kino, T., H. Hatanaka, M. Hashimoto, M. Nishiyama, T. Goto, M. Okuhara, M. Kohsaka, H. Aoki, and H. Imanaka. 1987. FK-506, a novel immunosuppressant isolated from a streptomyces. *J. Antibiotics* 40:1249–1255.

Nikkei Biotech, ed. 1989. *Nikkei Biotechnology Yearbook 89/90*. Nikkei Biotech Press, Tokyo.

Nukumi, M., F. Iwatani, M. Ohkuma, and Y. Odawara. 1987. Automatic colony transfer and analyzing system. *Hitachi Hyoron* 69:313–319.

Ohkawa, T., M. Nukumi, S. Nishimura, T. Okuda, T. Furumai, and T. Honda. 1989. Instrument for the Liquid Culture. Japanese Patent Kokai H1-225476 Sept. 8, 1989.

Umezawa, H., T. Aoyagi, H. Suda, M. Hamada, and T. Takeuchi. 1976. Bestatin, an inhibitor of aminopeptidase B, produced by actinomycetes. *J. Antibiotics* **29**:97–99.

Umezawa, H., T. Aoyagi, T. Hazato, K. Uotani, F. Kojima, M. Hamada, and T. Takeuchi. 1978. Esterastin, an inhibitor of esterase, produced by actinomycetes. *J. Antibiotics* **31**:639–641.

Umezawa, H., T. Aoyagi, K. Uotani, M. Hamada, T. Takeuchi, and S. Takahashi. 1980. Ebelactone, an inhibitor of esterase, produced by actinomycetes. *J. Antibiotics* **33**:1594–1596.

Umezawa, H., T. Aoyagi, S. Ohuchi, A. Okuyama, H. Suda, T. Takita, M. Hamada, and T. Takeuchi. 1983. Arphamenine A and B, new inhibitors of aminopeptidase B, produced by bacteria. *J. Antibiotics* **36**:1572–1575.

7

DNA Probes for the Identification of Pathogenic Foodborne Bacteria and Viruses

Keith A. Lampel, Peter Feng, and Walter E. Hill

Foodborne disease in the United States costs billions of dollars annually because of increased morbidity and mortality, time lost in the workplace, and reduced productivity (Archer and Kvenberg 1985). Because outbreaks of foodborne illnesses may be underreported by as much as a factor of 30 (Hauschild and Bryan 1980), the number of cases of gastroenteritis associated with food was estimated to be between 68 million and 275 million per year (Archer and Kvenberg 1985). Even at the lower end of this range, foodborne disease would be a major public health problem. From 1983 through 1987, 2397 outbreaks (54,453 cases) of foodborne illness were reported to the Centers for Disease Control (Bean et al. 1990). Of these, the causative agent was confirmed in only 910 cases (38%).

To reduce the adverse impact on public health caused by foodborne microbial disease, we must have methods to identify foodborne pathogens quickly. During the past decade, several rapid methods for the identification of foodborne bacteria have been developed. Many of these tests require pure cultures to determine the physiological, biochemical, or immunological characteristics of the bacteria; however, such methods usually cannot identify pathogenic strains. Research on bacterial virulence factors is now providing information on the mechanisms by which some of these bacteria cause disease in human hosts. It is possible, then, that tests may be devised to determine whether the bacteria are pathogenic, based on the genetic information they carry.

Bacteria that are the principal causes of foodborne disease include strains of *Bacillus cereus, Campylobacter jejuni, Clostridium botulinum, Clostridium perfringens, Escherichia coli, Listeria monocytogenes, Salmonella* spp., *Staphylococcus aureus, Shigella* spp., *Vibrio cholerae, Vibrio para-haemolyticus, Vibrio vulnificus, Yersinia enterocolitica,* and *Yersinia pseu-dotuberculosis* (Doyle 1985; Archer and Young 1988; Doyle 1989; Ryster and Marth 1989). Gene probes have been developed to identify these organisms as well as a number of viruses that cause foodborne disease.

GENE PROBES: DEVELOPMENT AND USE

Gene probes are nucleic acid molecules (usually DNA) of known genetic specificity. They are labeled to be easily identified, and are used to determine the presence of specific nucleotide sequences. How probes are developed and used, as well as technical considerations regarding labels, sensitivity, and specificity, are the major topics of this chapter.

How Probes Work

DNA hybridization has been known for more than 20 years (McCarthy 1967; Britten and Kohne 1968), but only within the last decade have advances in recombinant DNA techniques made it possible to prepare significant quantities of specific DNA fragments for use as gene probes. The key features of the DNA molecule that permit probes to function are complementarity of the two strands of polynucleotides that constitute the double helix and the sequence of nucleotides that encode genetic information. Two organisms that are closely related in evolutionary history will share many similar nucleotide sequences.

The strands of the DNA double helix can be separated by heat or alkaline treatment. If the strands of the probe DNA and the target DNA are complementary, they will form a double-stranded molecule when cooled or when the pH is lowered. Double helices composed of strands from different sources are designated as *hybrids;* the process of their formation is called *DNA hybridization.* Hybrids may also be formed between strands of RNA and DNA. To favor probe–target hybrids, the amount of probe should be in excess of the amount of target.

Development of Probes

Before a gene probe is developed, a specific target must be selected, for example, a particular bacterial genus in a food or a certain pathogenic strain. Gene probes for identifying an entire taxonomic group are often targeted to evolutionarily conserved genes, such as those encoding ribosomal RNA, whereas to identify a particular pathogenic strain, the usual target is a variant gene that encodes a virulence factor such as a toxin.

Therefore, much of the developmental work for probes may involve determining ribosomal RNA gene nucleotide sequences or discovering the genetic mechanisms of bacterial pathogenesis. Once the particular target has been selected, appropriate gene probes can be prepared either by purification of specific DNA fragments from genetically engineered bacterial strains or by synthesis (Gatt 1984; Caruthers 1985; Miyada, Studencki, and Wallace 1987). Both types of probes have been used to identify pathogenic strains in foods (Hill, Payne, and Aulisio 1983; Hill et al. 1985). Several reviews describe the development and use of gene probes in considerable detail (Arrand 1985; Matthews and Kricka 1988; Keller and Manak 1989).

Probe Test Formats

A critical step in conducting DNA hybridization assays using gene probes is the removal of labeled probe molecules that have not re-formed a double helix with the nucleic acid from the target. Two classes of techniques are used to solve this problem: solid phase, heterogeneous systems (Meinkoth and Wahl 1984), and homogeneous liquid hybridization assays.

For the colony hybridization assay, food homogenates are spread onto an appropriate medium. A replica of the resulting colonial array is made by gently pressing a solid support (usually a membrane of filter paper) onto the surface of the colonial growth, causing the colonies to adhere to the support. Cells can be lysed in situ by alkaline treatment (Grünstein and Hogness 1975), steam (Maas 1983), or microwave irradiation (Datta, Wentz, and Hill 1987). After hybridization, probe molecules remaining in solution can be easily removed by washing. A colony will remain labeled if it contains the same gene sequence as the probe. The label is determined by a system appropriate for that particular label. To calculate the number of cells in the original target that hybridized with the probe, the number of reacting colonies is multiplied by the dilution factor used to make the spread plate. Colony hybridization, then, is a quantitative method for determining the numbers of specific pathogens in a food.

In hybridization assays in which both probe and target DNAs are in solution, removal of unhybridized probe molecules is more difficult. Techniques such as the differential binding properties of double- and single-stranded DNA to hydroxylapatite (Britten and Kohne 1968) can be used; however, none have been developed for use with food. Recently, a convenient modification of a two-probe system was used to develop a rapid test for foods (Chan et al. 1989). The two probes are a "detector" probe labeled with a reporter group, fluorescein, that serves as an antigen, and a "capture" probe that binds to a specially coated dipstick. The bound detector probe can be removed from solution only if it is bound to the target nucleic acid, which, in turn, is linked to the dipstick by the capture probe. Horse-radish peroxidase-labeled antifluorescein antibodies are used for identification. This test is usually performed as a qualitative assay because cells must be grown in enrichment media.

PROBE PROS AND CONS

Genotype Versus Phenotype

Gene probe tests and those based on biochemical characteristics or immunological reactions have advantages and disadvantages. Gene probe assays are genotypic rather than phenotypic tests, that is, they can determine whether a gene is present, but not whether that gene is expressed. Because the environment of bacteria in vitro is much different from that of the human intestinal tract, some bacterial virulence genes may be expressed inefficiently and may give false-negative test results. On the other hand, a gene product (i.e., that which the gene encodes for) may be nonfunctional because of mutation, yet a positive probe reaction may occur.

Sensitivity

To provide reliable results, colony hybridization assays require colonies of about 10^5–10^6 cells. This is the number of target copies necessary to ensure the binding of a sufficient quantity of ^{32}P-labeled probe. High numbers of competing microflora in a food can interfere with the growth of the target organisms, and the minimum number of cells per colony required for the assay will not be reached. When the number of indigenous organisms reaches 10^5–10^6 per spread plate, a decrease in probe efficiency will result (Hill et al. 1985; Jagow and Hill 1986). The sensitivity of colony hybridization can be improved substantially by using selective media to limit the growth of competing flora (Jagow and Hill 1988).

Homogenous liquid hybridization systems have several advantages over the colony hybridization format. In colony hybridization, only 1–10% of the target nucleic acid affixed to a filter may be available to the probe because of hindrance by bacterial debris and entrapment within the solid support. However, in a liquid system, probe and target nucleic acids are free in solution, and both hybridization kinetics and target availability are improved. In addition to the characteristics that allow development of genus and sometimes species-specific probes, ribosomal RNA targets can provide more than 10,000 copies per cell, thereby greatly increasing the sensitivity of the probe.

Labels

Because of its sensitivity, ^{32}P is the most widely used probe label. However, because of the drawbacks associated with radioactivity (waste disposal, exposure of personnel), widespread acceptance and use of labeled gene probe methods will not occur until nonisotopic labels with a sensitivity and background comparable to those of ^{32}P are readily available.

Recently, a number of probe tests that use nonradioactive labels were reported. Using filtered foods and a biotinylated probe, Dovey and Towner (1989) observed 10^4 cells per dot blot, or the equivalent of 2×10^7 bacterial cells per gram of food. Their method, however, is not as sensitive as colony hybridization, which can determine the presence of 10^2 to 10^3 cells per gram of food. Romick, Lindsay, and Busta (1989) used the labeling method of Renz and Kurz (1985) to prepare a horseradish peroxidase-labeled probe to identify enterotoxingenic *E. coli* in foods by colony hybridization. Preliminary results with a digoxigenin-labeled cloned DNA fragment are encouraging; backgrounds appear to be lower than those that use biotinylated DNA (Riley and Caffrey 1990; Sommerfelt, Grewal, and Bhan 1990). Tests to determine the utility of nonradioactive labels for the analysis of foods by colony hybridization are not yet complete.

Specificity

The specificity of gene probes depends on the uniqueness of the target nucleotide sequence and the conditions under which the hybridization is conducted. Conserved regions of certain genes, such as ribosomal RNA genes, occur in several genera.

Factors that determine hybridization specificity include temperature and salt concentration. When the temperature is too low (low stringency), the probe may bind to target sequences that are not perfectly complementary, resulting in false-positive results. When the hybridization temperature is too high, exactly matched sequences will not remain double-stranded, and false-negative reactions will result. High stringency conditions are those under which only perfectly or very closely matched sequences will hybridize.

Specificity is also affected by the length of the DNA probe used. With cloned DNA fragments (usually between 100 and 2000 base pairs in length), minor mismatches do not occur, and a stringent hybridization might mean only 50–80% similarity between probe and target sequence. Because synthetic oligonucleotide probes are much shorter (15–30 nucleotides), they can be used with greater specificity to target small regions. Conditions can often be controlled precisely so that single base-pair mismatches are found (Wallace et al. 1979; Wood et al. 1985; Miller and Barnes 1986; Jacobs et al. 1988; Cebula and Koch 1990).

Viable Versus Nonviable Cells

Colony hybridization is useful for identifying microorganisms that produce disease by infection rather than by intoxication. In the latter instance, disease is usually caused by the ingestion of a preformed toxin that is present in the food, such as occurs in illnesses caused by *Staphylococcus aureus* and *Clostridium botulinum*. Viable cells may not be present in such foods, and gene probe results may be negative. However, with the advent of target DNA amplification techniques, viable cells are not needed to demonstrate the presence of toxin-producing cells.

AMPLIFICATION OF TARGETS

For most colony hybridization applications about 10^5-10^6 copies of the target nucleic acid are needed. Because the number of pathogenic microorganisms in foods may be considerably below this level, bacteria must be grown in cultures before probe testing. Recently, however, several techniques have been developed for amplifying particular segments of nucleic acid molecules (Saiki et al. 1988; Erlich, Gibbs, and Kazazian 1989; Guatelli et al. 1990; Kramer and Lizardi 1989). In less than 2 hours these in vitro replication techniques can produce over a millionfold increase in the amount of a specified target nucleic acid sequence, making it possible to determine the presence of a single bacterial cell in a food. These methods will considerably reduce the amount of time required to conduct hybridization tests by omitting the time-consuming and not always successful propagation of bacteria in culture.

One amplification method, the polymerase chain reaction (PCR), has already been applied to the testing of foods (Lampel et al. 1990; Bessesen et al. 1990), with gel electrophoresis used to determine the presence of target sequences. However, because these techniques are not dependent on the presence of viable cells, they may be of limited value when dead cells are present, as in pasteurized products. Amplification tests may also be used to identify foodborne viruses, such as rotaviruses, that are difficult to culture in the laboratory (Gouvea et al. 1990). Recently, bacterial 16S ribosomal RNA has been used as a template for the PCR (Wilson, Blitchington, and Greene 1990). This technique may be applicable to identify viable but nonculturable foodborne pathogenic bacteria.

TO PROBE OR NOT TO PROBE?

It is not easy to decide when a gene probe test is the most effective way to proceed in an analysis. Although probes can be useful as screening tests to identify particular bacterial species or strains, different probes are required to identify each characteristic. In selecting probes, the food type and the pathogens most likely to be associated with disease must be considered. Because probe tests can be labor-intensive, their use is efficient only when a number of foods are processed simultaneously.

Once a probe reaction signals the presence of a potential pathogen, subsequent confirmation should be carried out with a method based on a different principle, such as a biochemical, immunological, or animal model assay. As mentioned before, gene probes do not indicate pathogenicity, and because bacterial virulence is usually a multigene phenomenon, a biological assay may be the most accurate and appropriate confirmatory test.

Time

The choice of tests may be constrained by how soon the results are needed. Colony hybridization often requires three overnight periods (bacterial growth, hybridization, and autoradiography), and results may not be available until the third day. In some cases, hybridization and exposure time may be shortened to about 4 hours each. Liquid hybridization tests (Flowers et al. 1987; Curiale, Klatt, and Mozola 1990) usually require a 2-day, two-step enrichment, but hybridization and identification can be completed in 3 to 4 hours. When sensitivity, specificity, and cost of tests are equal, it makes no sense to select a rapid method for one aspect of a microbiological analysis if another aspect that takes longer must be carried out.

PROBE DETECTION OF FOODBORNE BACTERIAL AND VIRAL PATHOGENS

Of the 601 confirmed outbreaks of foodborne disease from 1983 through 1987, about 425 (71%) were due to bacterial infections rather than intoxications (Table 7-1). These bacteria were responsible for about 42,500 (78%)

TABLE 7-1   Agents Responsible for Confirmed Outbreaks of Foodborne Disease in the United States from 1983 to 1987

Agent	Outbreaks	Percent	Cases	Percent
Bacteria	601	66.0	50,217	92.2
Chemicals	232	25.5	1,244	2.3
Parasites	36	4.0	203	0.4
Viruses	41	4.5	2,789	5.1
Total	910	100	54,453	100

Source: From Bean et al. 1990.

of the cases. Clearly, rapid methods to identify the pathogenic strains of these bacterial species are necessary.

Probes for identifying foodborne pathogens were initially developed for use with clinical specimens (Tenover 1988). However, these probes work well for foods if suitable methods for test sample preparation are available. Some of the probes used to identify foodborne *E. coli, S. aureus, C. perfringens, C. botulinum, Salmonella,* and *Listeria* have been reviewed (Wernars and Notermans 1990). Some foodborne microorganisms and the probes used to identify them are listed in the following sections.

Bacillus cereus

The ubiquitous foodborne organism *Bacillus cereus* has been associated with several types of food, including rice, meats, desserts, spices, vegetables, and egg and dairy products. It is responsible for a major proportion of spoilage of dairy products, particularly milk (Ahmed, Moustafa, and Marth 1983; Billing and Cuthbert 1958; Coghill and Juffs 1979; Overcast and Atmaram 1974; Davies and Wilkinson 1973). Although *B. cereus* is found frequently in dairy products (Ahmed, Moustafa, and Marth 1983; Wong, Chang, and Fan 1988; Christiansson et al. 1989), few cases of food poisoning caused by *B. cereus* in these products have been reported (Johnsson 1984). Foodborne illnesses caused by *B. cereus* account for up to 23% of reported bacteria-related outbreaks (Kramer and Gilbert 1989).

Two distinct types of food poisoning caused by *B. cereus* (diarrheal and emetic) are induced by different toxins (Turnbull et al. 1979; Spira and Goepfert 1975). These illnesses are also distinguished by the time of onset: 8 to 16 hours for a delayed diarrheal syndrome, and 1 to 5 hours for an emetic response. Other presumptive virulence factors include phospholipase C complex (Otnaess et al. 1977) and sphingomyelinase (Ikezawa et al. 1978), now referred to as cereolysin AB (Gilmore et al. 1989), phosphatidylinositol phospholipase (Ikezawa et al. 1976), cereolysin (Colwell, Grushoff-Kosyk, and Bernheimer 1976), and cytolysin (Coolbaugh and Williams 1978).

Kramer and Gilbert (1989) have reviewed several microbiological and serological methods currently used to identify and isolate foodborne *B. cereus*. In one report, plasmid profiles were compared to serotyping and enterotoxin production of isolates associated with a food poisoning outbreak (DeBuono et al. 1988). The use of plasmid profiles was helpful in an outbreak where all isolates had the same plasmid profile; however, *B. cereus* strains of other outbreaks did not have the same profile.

At present there are no published reports of gene probes being used to identify virulent strains of foodborne *B. cereus*. If such probes are to be effective, several features of gram-positive spore-forming bacteria must be considered. Direct plating on selective enrichment would have to be performed in a most probable number (MPN) format; if colony hybridization is used, the difficulties of lysing gram-positive bacteria in situ must be overcome, and if spores are present in the food, their germination must be taken into account.

Campylobacter

Within the past decade, *Campylobacter jejuni* has been recognized as an important cause of gastroenteritis in humans and isolated from about 5% of individuals with diarrhea (Skirrow 1977). Epidemiological studies have implicated food as the major source of transmission to humans; raw milk (Oosterom et al. 1983) and undercooked poultry (Finch and Blake 1985) are the most important vehicles. Because the infective dose may be as low as 2 to 5 cells/ml (Robinson 1981), methods must be sensitive.

Many isolates associated with disease in humans elaborate a cholera-like enterotoxin (Ruiz-Palacios et al. 1983; Klipstein and Engert 1984) reported to be plasmid-encoded, but found on the chromosome as well (Lee, McCardell, and Guerry 1985). This enterotoxin may share nucleotide sequence similarity with the heat-labile enterotoxin of *E. coli* (Olsvik et al. 1984; Baig et al. 1986; Calva et al. 1989). A number of animal models have been tested to assess the virulence of *C. jejuni* (Ruiz-Palacios, Escamilla, and Torres 1981; Blaser et al. 1983; Caldwell et al. 1983), but none have proved reliable enough to enable researchers to identify the genetic basis by which *Campylobacter* causes disease.

Most gene probes for *Campylobacter* have sought to identify an entire taxonomic group rather than target a particular gene involved in pathogenesis. Romaniuk and Trust (1989) used oligonucleotide probes to 16S ribosomal RNA to identify campylobacter species. Also using 16S ribosomal RNA as a target, Martinelli, Carroll, and Donahue (1990) developed a quantitative chemiluminescent two-gene probe assay capable of identifying 10,000 cells of *C. jejuni*. A 16S ribosomal RNA probe specific for *Campylobacter pylori* (Morotomi et al. 1989) and a probe for ribosomal RNA of the thermophilic campylobacters, *C. jejuni, C. coli,* and *C. laridis* (Tenover et al. 1990), have been reported. A commercially available alkaline phosphatase-labeled oligonucleotide probe for identifying *C. jejuni* and *C. coli* was tested on pure cultures from clinical specimens, but may lack sensitivity when applied directly to stools (Olive, Johny, and Sethi 1990).

Picken, Wang, and Yang (1987) screened random restriction endonuclease-generated fragments and found three that are specific for *C. jejuni* and do not cross-react with *C. coli*. A 6.1-kilobase (kb) DNA fragment of unknown function from a genomic library of *C. jejuni* was specific for this species (Korolik, Cole, and Krishnapillai 1988). Using a nonisotopically labeled genomic DNA probe, Chevrier et al. (1989) developed a quantitative test to distinguish several *Campylobacter* species; however, this work was conducted with pure cultures and may not be applicable to foods.

Clostridium

Clostridium botulinum is a spore-forming anaerobe that may produce one or more of seven (types A through G) immunologically distinct neurotoxins. Although considerable work has been done to develop sensitive immunological methods to identify these highly potent and lethal toxins, gene probes have not yet been used to identify them in foods, perhaps because of the difficulty in obtaining authorization to use recombinant DNA techniques to clone and sequence *C. botulinum* genes for the acetylcholine-releasing inhibitors. Recently, the genes for toxins A, C1, and D have been sequenced (Thompson et al. 1990; Hauser et al. 1990; Binz et al. 1990), but probe development has not been reported. The availability of serotype-specific probes and PCR primers should allow rapid identification of toxin type without the need to produce the toxin itself or to recover viable cells or spores from contaminated foods.

   C. perfringens, commonly found in soil, is a resident of the intestinal tract and has been found in virtually all animals, including humans (Labbe 1989). Meat and poultry are the most common vehicles, but because the disease caused by this microorganism is relatively mild, many cases are probably not reported. Saito (1990) found *C. perfringens* in 46 of 80 (57%) stools taken from professional food handlers; however, a latex agglutination test revealed that only 7 of 55 isolated strains (13%) were positive for endotoxin production.

   The gene encoding a *C. perfringens* enterotoxin (CPE) has been cloned and sequenced (Van Damme-Jongsten et al. 1990). Probes have been constructed and tested on strains isolated from foods, feces, and the environment, but all were related to cases of foodborne diseases. Of 53 foodborne isolates tested, 35 (66%) reacted with the CPE gene probes. Isolates from fecal specimens showed a 57% (110/192) positive rate. None of the four environmental specimens tested reacted with the CPE gene probes.

Escherichia coli

Four classes of diarrheagenic *Escherichia coli* are associated with foodborne disease in humans (Levine et al. 1987):

- Enteropathogenic (EPEC)
- Enterotoxigenic (ETEC)
- Enterohaemorrhagic (EHEC)
- Enteroinvasive (EIEC)

These pathogens have been grouped on the basis of their clinical manifestations, epidemiology, serotype as defined by O (somatic) and H (flagellar) antigens, invasive properties, and adherence to intestinal cells. Although probes for the identification of pathogenic *E. coli* were first developed for testing clinical or environmental specimens, they are the most widely used probes for identifying virulent strains of bacteria in foods (Hill et al. 1983; Hill, Payne, and Aulisio 1983; Hill and Payne 1984; Ferreira et al. 1986). After these genes were sequenced, oligonucleotide probes were constructed and shown to be equally effective for identifying enterotoxigenic *E. coli* strains in pure culture or contaminated foods (Hill et al. 1985; 1986).

ETEC

The pathogenicity of ETEC is dependent upon three factors: ingestion of a sufficient number of bacteria, the presence of plasmid-encoded heat-labile enterotoxin (LT) or a heat-stable enterotoxin (ST), and production of colonization factors essential for adherence to the mucosal layer of the small intestine. These stains are often implicated as the causative agent of infant diarrhea in Third World nations (Black et al. 1982)

DNA probes for virulence genes expressing products for ST and LT enterotoxins, invasion, Shiga-like toxin (SLT) or verotoxin (VT), and adherence have been developed. Several of the genes encoding LT and ST enterotoxins have been used as gene probes (Echeverria et al. 1986; Maas et al. 1985; Moseley et al. 1980; 1982). In these studies, DNA fragments from the LT and ST genes were isolated, labeled with ^{32}P, and used in colony hybridization blots. Recently, nonradioactive probes have been developed. Probes derived from enterotoxin genes were alkaline phosphatase-conjugated (Seriwatana et al. 1987), biotinylated (Bialkowska-Hobrzanska 1987; Kirii et al. 1987), or digoxigenin-labeled (Riley and Caffrey 1990; Sommerfelt, Grewal, and Bhan 1990). In all these studies, either pure cultures or clinical specimens were used to assay the efficacy of the probes with the nonisotopic label.

EPEC

The EPEC strains are distinguishable from other *E. coli* pathogens by their ability to adhere to HEp-2 cells in vitro (Cravioto et al. 1979) and by O serotype. They do not produce ST or LT, and their mode of invasion is unlike that of EIEC (Levine et al. 1978). This group has a plasmid that carries the genetic information for enteroadherence factors (EAF), for which a probe was developed (Nataro et al. 1985) and used in clinical situations (Echeverria et al. 1989).

EHEC (SLT-producing E. coli; SLTEC)

Pathogenic *E. coli* O157:H7 strains that produce SLT (or VT) cause hemorrhagic colitis and hemolytic uremic syndrome. There are two SLTs, SLT-I and SLT-II (O'Brien and Holmes 1987), and the genes for these cytotoxins reside on two lysogenic bacteriophages. Probes from these genes have been developed by Newland and Neill (1988). EHEC strains also have a plasmid that carries a gene for a fimbrial antigen necessary for attachment to epithelial cells (Levine et al. 1987). The DNA probes of Newland and Neill (1988) were tested in foods that were artificially contaminated with *E. coli* O157:H7 strains (Samadpour et al. 1990). The sensitivity of colony hybridization using these probes was 1.3 CFU per gram of food. These studies included an enrichment step before the bacteria were plated and filtered. Most foods that were not intentionally contaminated contained virtually no EHEC strains, except for an SLTEC strain found in surimi-based salad and in blueberries. This result indicates that all SLT-producing strains can be found with these probes, regardless of serotype. The time from enrichment to identification was 2 days with the dot blot method, but no isolate could be recovered; with colony hybridization, however, cells can be identified in 3 to 4 days and an isolate obtained for further study.

Salmonella

It is estimated that 2 million to 4 million cases of human salmonellosis occur annually in the United States. Most often, outbreaks can be linked to contaminated poultry, eggs, dairy products, pigs, and raw meats. The most common human pathogen of this genus is *Salmonella typhimurium,* although recently there has been increased concern regarding *S. enteritidis,* which is associated with eggs (St. Louis et al. 1988). The numbers of foodborne outbreaks caused by salmonellae throughout the United States (Chalker and Blaser 1988) and the world is increasing (D'Aoust 1989).

Salmonellae are classified biochemically and serologically into serovars (or antigenic species); an antigenic system (White 1926; Kauffmann 1966) defines more than 2000 types of *Salmonella.* These bacteria are one of several microbial pathogens that may be present during all stages of food and drink processing, from raw materials through transportation, storage, and consumer acquisition (Jackson, Langford, and Archer 1991). An individual with AIDS is 19.2 times more likely than a healthy person to contract salmonellosis (Celum et al. 1987).

A DNA hybridization method using several cloned DNA fragments that are unique to this genus was developed to identify *Salmonella* in foods (Fitts et al. 1983; Fitts 1985). The foods were spiked with several species of *Salmonella,* including *S. typhimurium* and incubated overnight in a nonselective medium; bacteria were collected on nitrocellulose filters by filtration and probed with the DNA fragments. Hybridization results showed that these DNA fragments were specific for the *Salmonella* species tested, and there was no cross-reactivity with background flora in the food.

A collaborative study compared the efficacy of DNA probes with standard microbiological procedures (Flowers 1985; Flowers et al. 1987). Several of the specific DNA fragments of Fitts et al. (1983) were used as probes in hybridization assays of salmonellae-inoculated foods. The efficacy of the DNA probes was as good as or better than the standard culture method, depending on the food, and was approved as official first action by the Association of Official Analytical Chemists (AOAC) in 1989.

Using flour or flour-based bakery mixes, Sall et al. (1988) compared cultural methods with a commercially available DNA hybridization test for *Salmonella*. In a 1-year study, they found the probe method to be more sensitive than microbiological methods. In one case, a lactose-positive *Salmonella* isolate was correctly identified by the probe analysis, but not by other methods. In another study (Izat et al. 1989), probes were tested in poultry isolates and showed no false-positive reactions.

Curiale, Klatt, and Mozola (1990) used salmonella-specific DNA probes targeted to ribosomal RNA in a hybridization assay without a radioisotope. The probes in this test hybridize to any *Salmonella* ribosomal RNA, excluding subgenus V salmonellae. This assay, which was tested in several foods and was specific for all salmonellae except seven isolates, has been recommended for adoption as official first action by the AOAC. A biotinylated DNA probe was also developed to identify *Salmonella* in foods (Gopo et al. 1988).

Shigella and Enteroinvasive Escherichia coli

Four *Shigella* spp. (*S. sonnei, S. flexneri, S. dysenteriae,* and *S. boydii*) and enteroinvasive *E. coli* (EIEC) are the causative agents of bacillary dysentery or shigellosis. From 1965 to 1988, cases in which *Shigella* spp. were isolated in the United States increased from 5.4 to 10.1 per 100,000 people (Centers for Disease Control 1990). In 1988, state health departments reported 22,796 cases to the Centers for Disease Control, the highest number reported since records were started in 1965.

Several factors make the rapid identification of an outbreak source critical in reducing the number of human infections. The estimated infective dose of *Shigella* is low—between one and 10 organisms (Bryan 1979)—and the spread can be linked to poor personal hygiene (Smith 1987), which leads to rapid dissemination of shigellae. In addition, the prophylactic use of antimicrobials is not recommended, even to prevent secondary infections, because of the rapid shift in resistance patterns in *Shigella.*

The sources of foodborne bacillary dysentery outbreaks are difficult to pinpoint. Standard microbiological methods (Mehlman 1984) to isolate *Shigella* spp. are cumbersome, protracted, and not always successful. Thus, the use of DNA probes to identify pathogenic *Shigella* and EIEC is an attractive alternative for minimizing the impact of these microorganisms on the public health.

The first gene probes (Boileau, d'Hauteville, and Sansonetti 1984; Small and Falkow 1986; Taylor et al. 1988; Wood et al. 1986) developed to identify EIEC and *Shigella* spp. were DNA fragments derived from the large invasion plasmid (Sansonetti, Kopecko, and Formal 1981) found in strains that can invade human colonic epithelial cells (LaBrec et al. 1964). Recently, specific DNA from the invasion plasmid antigen (*ipaH*) genes have been used as probes for EIEC and *Shigella* spp. (Venkatesan et al. 1988; 1989). However, these applications have been solely for clinical use. For example, a probe derived from the *ipaH* gene has been tested as a general identification probe. The *ipaH* gene is found on the *Shigella* chromosome as well as on the large invasion plasmid of virulent strains. Therefore, avirulent strains of *Shigella* can be identified by this probe. To increase the sensitivity of identifying EIEC and *Shigella* spp. in clinical isolates, oligonucleotides from the invasion plasmid have been used as primers for the PCR (Frankel et al. 1990). An alkaline phosphatase-conjugated oligonucleotide probe was used against clinically isolated strains of *Shigella* spp. and shown to be very effective in identifying Sereny-positive strains (Panda et al. 1990). However, these tests were performed only with isolated cultures.

A synthetic oligonucleotide probe derived from a 2.5-kb pair *Hind*III fragment from the invasion plasmid was found to be specific for EIEC and invasive *Shigellae,* and was used to identify these bacteria in foods (Jagow and Lampel 1989). However, the efficacy of this probe was dependent on the food tested, for example, *Shigella* was not found in artificially contaminated alfalfa sprouts because the background microbial flora outgrew the seeded *Shigella.*

In general, foodborne outbreaks of shigellosis are recognized only after confirmation by positive clinical specimens. Usually, an antibiotic profile of the isolate is obtained by the clinical laboratory. This information can be used to reduce competing microbial flora by incorporating antibiotics into a selective medium for isolating the suspected bacteria directly from the food (Shook, Jagow, Harrell, and Lampel, manuscript in preparation).

Staphylococcus aureus

Staphylococcus aureus produce at least five immunologically distinct but genetically related enterotoxins that are a major cause of food poisoning. About 8% of the confirmed outbreaks of foodborne disease are believed to be caused by the ingestion of preformed enterotoxins produced by this species (Table 7-2; Bean et al. 1990). A considerable amount of work has been done on the structure of some of these toxin proteins, and a number of the genes encoding them have been analyzed.

There is some controversy as to the location of *entA,* the gene encoding staphylococcal enterotoxin A (SEA). Originally, it was thought to be located chromosomally (Shafer and Iandolo 1978) and to have a specific map location (Mallonee, Glatz, and Pattee 1982). Other evidence suggests that the *entA* gene is a movable genetic element (Betley et al. 1984), and most recently it was encoded by a bacteriophage (Betley and Mekalanos 1985). The gene encoding SEA has been sequenced (Betley and Mekalanos 1988).

The gene for staphylococcal enterotoxin B (SEB) has been cloned (Ranelli et al. 1985) and sequenced (Jones and Kahn 1986), as has that for enterotoxin C_1 (Bohach and Schlievert 1987). These two genes are linked on one plasmid in one strain of *S. aureus* (Altboum, Hertman, and Sarid 1985). Notermans, Heuvelman, and Wernars (1988) developed synthetic probes that were specific for SEB and enterotoxin C (SEC), targeted to a region in the gene where the amino acid sequences of the toxins were identi-

TABLE 7-2   Bacterial Species Implicated in Confirmed Outbreaks of Foodborne Illness in the United States from 1983 to 1987

Bacterium	Outbreaks	Percent	Cases	Percent
Bacillus cereus	16	2.7	261	0.5
Brucella	2	<0.1	38	0.1
Campylobacter	28	4.7	700	1.4
Clostridium botulinum	74	12.3	140	0.3
C. perfringens	25	4.2	2,743	5.5
Escherichia coli	7	1.2	640	1.3
Salmonella	342	56.9	31,185	62.1
Shigella	44	7.3	9,971	19.9
Staphylococcus aureus	47	7.8	3,181	6.3
Streptococci, Group A	7	1.1	1,001	2.0
Streptococci, other	2	<0.1	85	0.2
Vibrio cholerae	1	<0.1	2	<0.1
V. parahaemolyticus	3	0.5	11	<0.1
Other	3	0.5	259	0.5
Total	601	100	50,217	100

Source: From Bean et al. 1990.

cal (Huang and Bergdoll 1970; Schmidt and Spero 1983). The probe developed by Notermans, Heuvelman, and Wernars (1988) was used successfully by Trucksess et al. (1988) to find SEB-producing *S. aureus* in artificially contaminated crabmeat. The genes for enterotoxin D (SED) (Bayles and Iandolo 1989) and E (SEE) (Couch, Soltis, and Betley 1988) have been cloned and sequenced. Even though the enterotoxin genes share 40–80% similarity, probes were developed to differentiate SEA, SEB, SEC, and toxic shock syndrome toxin 1 by targeting nonconserved regions (Neill et al. 1990).

Vibrio

Gene probes have been developed to identify disease-causing stains of three *Vibrio* species recognized as foodborne pathogens: *V. cholerae, V. parahaemolyticus,* and *V. vulnificus.* Although endemic in some parts of the world, *V. cholerae* causes only sporadic outbreaks in the United States. A major virulence factor, the cholera enterotoxin (CT), bears an evolutionary resemblance to the heat-labile enterotoxin (LT) of *E. coli* (Moseley and Falkow 1980). Synthetic probes to CT can differentiate between CT and LT (Hill and Payne, unpublished results; Cook, Wachsmuth, and Hill, submitted for publication). Some strains of *V. cholerae* produce a heat-stable enterotoxin related to that of *E. coli,* and oligonucleotide probes have been used to identify the gene for this toxin (Hoge et al. 1990).

Gastroenteritis caused by *V. parahaemolyticus* is associated with the consumption of seafood. Although the role of a thermostable direct hemolysin in disease is not clearly understood, this molecule is elaborated mostly by clinical isolates, but only by a few environmental isolates (Miyamoto et al. 1969). The entire hemolysin gene has been used as a probe in colony hybridizations (Nishibuchi et al. 1985). The gene has also been sequenced (Nishibuchi and Kaper 1985). Synthetic DNA probes were specific for hemolytic strains when high stringency was used (Nishibuchi et al. 1986).

Illness caused by *V. vulnificus* has been associated with exposure to seawater and the consumption of raw shellfish; the mortality rate is about 50% (Tacket, Brenner, and Blake 1984). A cloned fragment containing a cytotoxin-hemolysin gene is a species-specific gene probe (Wright et al. 1985) that was used to identify *V. vulnificus* in oysters (Morris et al. 1987); this region has been sequenced (Yamamoto et al. 1990) and oligonucleotide probes have been synthesized. *V. vulnificus* was identified in artificially contaminated oysters by a PCR used to amplify a region of this gene (Hill et al. 1990).

Yersinia

Infections caused by *Yersinia enterocolitica* and *Y. pseudotuberculosis* commonly occur in the form of gastroenteritis in humans. The characteristic symptoms are fever and abdominal pain with diarrhea; nausea and vomiting occur occasionally (Archer and Young 1988; Schiemann 1989). Enteritis caused by *Yersinia* spp. is not a frequent cause of human infection in the United States; an estimated 3000 to 20,000 cases occur each year (Roberts 1989; Todd 1989). In the past 15 years, four outbreaks of yersinosis caused by *Y. enterocolitica* serotype O:8 were traced to contaminated chocolate milk, milk powder, and water used in the preparation of tofu (Aulisio et al. 1983). An outbreak caused by serotype O:13 was also traced to pasteurized milk, but contamination of the outside of the milk containers was probably the cause of human illness (Aulisio, Lanier, and Chappel 1982; Roberts 1989). In Japan, 12 outbreaks over the past 13 years were caused by *Y. pseudotuberculosis,* mostly serotype 4b (Tsubokura et al. 1989). Sporadic infections by *Y. enterocolitica* serotypes O:3 and O:9 are also a common problem in Northern Europe and Scandinavia, but few *Y. enterocolitica* outbreaks have been reported in the United States (Archer and Young 1988; Cover and Aber 1989; Doyle 1986; Schiemann 1989). *Yersinia* spp. are found in water, soil, vegetables, milk, and a wide variety of meats, but the only known reservoir for *Y. enterocolitica* is pork (Cover and Aber 1989; Mafu et al. 1989; McManus and Lanier 1987).

Isolation of *Yersinia* from foods is a lengthy procedure. The FDA *Bacteriological Analytical Manual* specifies cold enrichment in broth for 10 days at 10°C followed by alkali treatment and plating on selective medium. Few *Yersinia* spp. encountered in foods are virulent; therefore, isolates must also be tested for pathogenicity (Doyle 1986; Hill, Payne, and Aulisio 1983; Schiemann 1989). Virulence and virulence-related phenotypes in *Yersinia* are associated with the presence of a 70-kb plasmid (Doyle 1986; Kay, Wachsmuth, and Gemski 1982; Portnoy, Moseley, and Falkow 1981; Schiemann 1989); however, chromosomal genes are also required for full virulence (Doyle 1986; Isberg, Voorhis and Falkow 1987; Kay, Wachsmuth, and Gemski 1982; Miller et al. 1989; Portnoy, Moseley, and Falkow 1981; Schiemann 1989).

The use of DNA probes to identify yersiniae in foods was first reported by Hill, Payne, and Aulisio (1983), who used a pool of 3 *Bam*HI fragments cloned from the Ca^{++} dependency region of the virulence plasmid (Portnoy et al. 1984). Colony hybridization analyses of 11 artificially contaminated foods showed probe efficiencies from 66% to 100% (average 86%) (Jagow and Hill 1986). The sensitivity of the probe assay was affected by the number of bacteria present in the food. When the background exceeded 10^8 cells per gram of food, a selective enrichment consisting of a brief treatment with a 1:25 dilution of 0.5% KOH–0.5% NaCl was needed to improve the sensitivity (Jagow and Hill 1988). It was estimated that one yersinia cell in the presence of 3×10^5 background counts can be identified in 24–48 hours at 26°C (Jagow and Hill 1986). In later modifications of the assay a single DNA fragment was used, and the nitrocellulose filter was replaced with the more economical Whatman 541 membranes (Jagow and Hill 1988). Probe identification of virulent *Yersinia* correlated well with the pathogenicity test results obtained with autoagglutination and suckling mouse lethality assays (Hill, Payne, and Aulisio 1983).

A 2.6-kb *Bam*HI fragment cloned from the virulent plasmid of *Y. enterocolitica* was also found to be highly conserved and specific for pathogenic yersiniae (Miliotis et al. 1990; Robins-Browne et al. 1989). This fragment encodes genes associated with the production of conjunctivitis in guinea pigs and with cytotoxicity for HEp-2 cells in vitro (Miliotis et al. 1989, 1990). A 24-base oligonucleotide (SP12) to this region was evaluated as a probe for the analysis of pathogenic *Y. enterocolitica* in seven types of seeded foods. The probe identified 10^4–10^6 yersiniae per gram of food in the presence of background bacteria (Miliotis et al. 1990).

The use of another oligonucleotide probe to identify yersiniae in foods was reported by Kapperud et al. (1990). The 19-base plasmid-specific probe is directed to the *yopA* gene, which encodes a temperature-inducible outer membrane protein. Analysis of three seeded foods by colony hybridization showed that probe efficiency ranged from 33% to 82% in the presence of 3×10^4 cells per gram of background bacteria. No selective enrichment was used in any of these probe assays; however, an enrichment step may improve efficiency (Jagow and Hill 1988).

All DNA probes discussed previously are targeted to the conserved regions of the virulence plasmid, which is homologous among the various pathogenic *Yersinia* species (Portnoy, Moseley, and Falkow 1981). These plasmid-specific probes can therefore identify, but cannot speciate pathogenic *Yersinia*. A further disadvantage is that the virulence plasmid in yersiniae is not stable (Schiemann 1989); hence, strains that may have lost the plasmid during isolation will not be found by these probes. Decreased probe sensitivity caused by plasmid loss may be resolved by using probes directed to chromosomal virulence markers that are stably inherited. The ability of *Yersinia* to invade mammalian tissue cells is encoded on the chromosome by the invasive gene (*inv*) in *Y. pseudotuberculosis* (Isberg, Voorhis, and Falkow 1987) and by the *inv* and attachment invasion locus gene (*ail*) in *Y. enterocolitica* (Miller et al. 1989).

Several probes specific for the invasion genes of *Yersinia* have been characterized. The *inv* gene of *Y. pseudotuberculosis* is 3.2 kb and encodes for invasin, a protein that enables yersiniae to invade mammalian cells (Isberg, Voorhis, and Falkow 1987). A 4.5-kb *Bam*HI probe carrying the entire *Y. pseudotuberculosis inv* gene, including flanking regions, hybridized only with *Y. pseudotuberculosis* and *Y. pestis* isolates (Robins-Browne et al. 1989). Although only the former species causes gastroenteritis, both pathogens share more than 90% DNA homology overall (Brenner et al. 1976). Similarity, a 21-base *inv*-specific oligonucleotide probe (INV-3) was specific only for *Y. pseudotuberculosis*. All INV-3 probe-reactive isolates were invasive for HeLa cells (Feng 1990). A third probe directed to the *inv* gene of *Y. pseudotuberculosis* is a 2.4-kb *Cla*I-*Xho*I fragment that contains only internal sequences of the *inv* gene (Miller et al. 1989). Hybridization studies showed that the internal probe hybridized with both *Y. pseudotuberculosis* and *Y. enterocolitica,* but not with other bacteria. The *inv* genes from these species share some sequence similarity (Feng 1990; Miller et al. 1989).

Only one probe directed to the *inv* gene of *Y. enterocolitica* has been evaluated (Miller et al. 1989). This probe, a 3.6-kb *Cla*I fragment, hybridized with all yersiniae, including virulent or avirulent isolates (Isberg, Voorhis, and Falkow 1987; Miller et al. 1989). Two probes are specific for *ail,* the other invasion gene on the *Y. enterocolitica* chromosome. A 1.2-kb *Ava*I-*Cla*I probe carrying the entire *ail* gene (0.65 kb) identified only pathogenic *Y. enterocolitica* (Miller et al. 1989; Robins-Browne et al. 1989). Similarly, an oligonucleotide probe (PF-13) directed to *ail* hybridized only with *Y. enterocolitica* isolates that were HeLa cell-invasive (Feng 1990). The *ail* gene probes, therefore, may be useful to detect pathogenic isolates of *Y. enterocolitica.*

Although several probes directed to the chromosomal virulence markers in *Yersinia* have been characterized, none have been tested for identifying yersinia in foods. All the probes discussed in this section are summarized in Table 7-3.

Viruses

DNA probes are being used increasingly to identify viruses in clinical specimens, but their use with foods has been limited. Enteroviruses are the group of viruses most commonly isolated from foods (Cliver, Ellender, and Sobsey 1984). This group has single-stranded RNA and contains more than 70 serotypes (Larkin 1986; Rotbart, Levin, and Villarreal 1984). Common methods used to identify enteroviruses include isolation from tissue cultures and seroneutralization, which often require more than 2 weeks to complete and are limited, since not all enteroviruses can be cultured in vitro (Petitjean et al. 1990; Rotbart, Levin, and Villarreal 1984).

TABLE 7-3   Probes Used to Detect Foodborne Pathogenic Bacteria

Microorganisms	Target[a]	Probe Type[b]	Selected References(s)[c]
Campylobacter	Mixed fragments	C	Rashtchian and Curiale 1989
spp.	Ribosomal RNA	C	Rashtchian, Abbott, and Shaffer 1987
C. jejuni	Ribosomal RNA	O	Stolzenbach et al. 1988
	Ribosomal RNA	O	Olive, Johny, and Sethi 1990
	Ribosomal RNA	O	Martinelli, Carroll, and Donahue 1990
Clostridium spp.			
C. botulinum	Neurotoxin A	S	Thompson et al. 1990
	Neurotoxin C1	S	Hauser et al. 1990
	Neurotoxin D	S	Binz et al. 1990
C. perfringens	CPE	O	Van Damme-Jongsten et al. 1990
Escherichia coli	Ribosomal RNA	O	Hsu et al. 1989
	ST	C	Moseley et al. 1982
	ST	O	Lee et al. 1983
	ST	O	Hill et al. 1985
	LT	C	Moseley et al. 1980
	LT	O	Moseley et al. 1982
	LT	C	Hill et al. 1983
	LT	C	Guth et al. 1986
	SLT	C	Smith et al. 1987
	SLT	C	Newland and Neill 1988
	SLT	C	Levine et al. 1987
	SLT	C	Brown et al. 1989
	SLT	C and O	Sowers et al. 1989
Salmonella spp.	Unknown	C	Fitts et al. 1983
	Not reported	O	Scully, Chapis, and Kuritza 1989
	Ribosomal RNA	O	Curiale, Klatt, and Mozola 1990
S. enteritidis	Unknown	C	Tompkins et al. 1986
Shigella spp.	Plasmid	C	Boileau, d'Hauteville, and Sansonetti 1984
	Invasiveness	C	Small and Falkow 1986
	Invasiveness	O	Jagow and Lampel 1989
Staphylococcus	SEA	O	Betley and Mekalanos, 1988
aureus	SEB	C	Jones and Kahn 1986
	SEB	O	Neill et al. 1990
	SEB, SEC	O	Notermans, Heuvelman, and Wernars 1988
	SEA, SEB, SEC	O	Couch, Soltis, and Betley 1988
Vibrio spp.			
V. cholerae	CT	C	Kaper, Moseley, and Falkow 1981
V. parahaemo-	Hemolysin	C	Nishibuchi et al. 1985
lyticus	Hemolysin	O	Nishibuchi et al. 1986

TABLE 7-3   (continued)

Microorganisms	Target[a]	Probe Type[b]	Selected References(s)[c]
V. vulnificus	Cytotoxin-hemolysin	C	Wright et al. 1985
	Cytotoxin-hemolysin	C	Morris et al. 1987
	Cytotoxin-hemolysin	C	Kaysner et al. 1987
	Cytotoxin-hemolysin	O	Yamamoto et al. 1990
Y. enterocolitica	Plasmid	C	Portnoy, Moseley, and Falkow 1981
	Calcium dependency	C	Hill et al. 1983
	Calcium dependency	C	Jagow and Hill 1986
	Calcium dependency	C	Jagow and Hill 1988
	ail	C	Miller et al. 1989
	ail	O	Feng 1990
	Invasiveness	O	Miliotis et al. 1989
Y. pseudotuber-culosis	Invasiveness	C	Isberg, Voorhis, and Falkow 1987
	Invasiveness	O	Feng 1990

[a]*C. perfringens* enterotoxin; ST, heat-stable enterotoxin; LT, heat-labile enterotoxin; SLT, Shiga-like toxin; SEA, SEB, SEC, staphylococcal enterotoxins A, B, and C respectively; CT, cholera toxin; *ail*, attachment invasion locus.
[b]C, cloned; O, synthetic oligonucleotide; S, sequence only.
[c]Listing priority is given to references reporting the analysis of food, clinical, or environmental specimens rather than probe development, because the latter papers are usually cited when applications are described. If no probes have been reported, the cloning or sequencing paper is cited.

Among the enteroviral group, hepatitis A virus (HAV) is most often associated with food and water transmission (Cliver, Ellender, and Sobsey 1984; Petitjean et al. 1990; Rotbart, Levin, and Villarreal 1984) (Table 7-4). About 4000 cases of HAV were recorded in the United States between 1944 and 1982; however, only 5–10% of the cases are actually reported. Another source estimates the yearly cases of foodborne HAV in the United States to be 35,000 (Todd 1989). The food vehicle most often contaminated with HAV is raw or undercooked shellfish, with salads, sandwiches, fruit, milk, and pastries also implicated (Cliver, Ellender, and Sobsey 1984; Jiang et al. 1986; Jiang, Estes, and Metcalf 1987).

TABLE 7-4   Viruses Implicated in Confirmed Outbreaks of Foodborne Illness in the United States from 1983 to 1987

Virus	Outbreaks	Percent	Cases	Percent
Hepatitis A	29	70.7	1,067	38.3
Norwalk	10	24.4	1,164	41.7
Other	2	4.9	558	20.0
Total	41	100	2,789	100

Source: From Bean et al. 1990.

Several probes are available for identifying HAV in environmental specimens. The probes include a cDNA fragment specific for the 3' end of the HAV genome (Gerba, Margolin, and Trumper 1988); cDNA clone fragments that constitute 99% of the HAV genome (Jiang et al. 1986); and RNA probes generated from the cDNA fragments using reverse transcriptase (Jiang, Estes, and Metcalf 1987). Analysis of purified viral particles generally showed good probe specificity and a sensitivity of 10^4 particles. In dot blot analysis of seeded environmental specimens, the efficiencies ranged from 19% to 74% within 48 hours, depending on the elution and extraction procedures used. Generally, isotopically labeled RNA probes were eight times more sensitive than cDNA probes or nonisotopic-labeled RNA probes, probably because of the inherent stability of RNA:RNA hybrids. The finding of one plaque-forming HAV unit within 72 hours from estuarine specimens has also been reported (Gerba, Margolin, and Trumper 1988). These probes have not been tested for identification of HAV in foods.

Probes potentially useful for identification of other enteroviruses in foods include cDNA clone fragment(s) from coxsackievirus B3 (Hyypia et al. 1984), coxsackievirus B4 (Chatterjee, Kaehler, and Deibel 1988), and poliovirus type 1 genome (Rotbart, Levin, and Villarreal 1984); an oligonucleotide probe from the 5' end of the picornavirus genome (Bruce et al. 1989); and RNA probes derived from genomic sequences of poliovirus (Petitjean et al. 1990). Probe sensitivities for identifying enteroviral RNA in infected tissue culture cells or cell lysates may be as high as 85%; however, for analysis of crude clinical specimens, sensitivities ranged only from 2.5 to 33%. In general, probe assays were more sensitive than seroneutralization for identifying enteroviruses (Bruce et al. 1989; Cukor and Blacklow 1984; Petitjean et al. 1990).

Norwalk agent and rotavirus are the most frequent causes of viral gastroenteritis in humans; the disease occurs endemically and epidemically worldwide (Cliver, Ellender and Sobsey 1984; Cukor and Blacklow 1984). Gastroenteritis caused by Norwalk agent has been traced to contaminated drinking water, raw and undercooked foods, vegetables, salads, milk, eggs, and various meats (Cukor and Blacklow 1984; Larkin 1981). In the United States, there are an estimated 180,000 cases of foodborne Norwalk virus infections yearly (Todd 1989), mostly from ingestion of raw shellfish (Cliver, Ellender, and Sobsey 1984; Larkin 1986). Because Norwalk agent has yet to be propagated in vitro (Cukor and Blacklow 1984), little information is available about the virus for definitive classification and probe development.

Rotavirus, the major cause of infant diarrhea, has not been associated with foodborne disease outbreaks (Cliver, Ellender, and Sobsey 1984). Rotavirus belongs to the family Reoviridae and carries 11 double-stranded RNA segments in its genome (Cukor and Blacklow 1984). Probes developed for rotaviruses include cDNA fragments, which can identify group B rotaviruses in fecal specimens (Eiden et al. 1989) and rotavirus serotypes (Zheng et al. 1989); RNA probes, with a sensitivity as low as 8 picograms (pg) (Flores et al. 1983); and a nonradioactive oligonucleotide with comparable sensitivities to three commercial enzyme immunoassays and the silver-stain polyacrylamide gel method used to identify rotaviruses (Arens and Swierkosz 1989).

Other cases of viral transmission through foods include poliomyelitis in unpasteurized or recontaminated milk, and echovirus 4 in cole slaw (Cliver, Ellender, and Sobsey 1984). These, however, are isolated incidents. Currently, probe assays may lack sufficient sensitivity to determine the presence of low numbers of viral particles in foods. The use of target amplification reactions and development of methods to directly extract viral nucleic acids from foods should improve sensitivity and realize the potential usefulness of probe assays in identifying foodborne viruses. The probes discussed in this section are summarized in the Table 7-5.

CONCLUSION

A considerable amount of research has been performed on the virulence factors of some of the important foodborne pathogens, and a number of

TABLE 7-5   Nucleic Acids Probes Used to Detect Viruses in Food and Water

Probe	Target	Reference
	Enteroviruses	
DNA	cDNA from coxsackievirus B3	Hyypia et al. 1984
DNA	cDNA from poliovirus type 1	Rotbart, Levin, and Villarreal 1984
DNA	cDNA from coxsackievirus B4	Chatterjee, Kaehler, and Deibel 1988
Oligo	5' end of picornavirus	Bruce et al. 1989
RNA	Poliovirus type 1	Petitjean et al. 1990
DNA	cDNA—99 percent of HAV genome	Jiang et al. 1986
RNA	cDNA of HAV genome	Jiang, Estes, and Metcalf 1987
DNA	cDNA from 3' end of HAV	Gerba, Margolin, and Trumper 1988
	Rotavirus	
RNA	In vitro transcription	Cukor and Blacklow 1984
DNA	cDNA from group b rotavirus	Eiden et al. 1989
DNA	cDNA from segment 9	Zheng et al. 1989
Oligo	Molecular Biosystems, Inc.	Arens and Swierkosz 1989

genes responsible for pathogenic determinants have been identified. Most of these genes have been cloned and sequenced, and gene probes have been constructed and tested. Probes based on ribosomal gene targets have been used successfully to identify particular taxonomic groups at the genus level. The development of probes was initiated to provide tools for testing clinical specimens, and only a few probes have been used to test foods for the presence of bacterial pathogens. Probes currently available could be used in about 70% of foodborne disease outbreaks. Probes for viruses may also prove to be extremely useful because other methods for identifying viruses in foods have not been routinely successful. Target amplification will increase the sensitivity of probe tests and can reduce the time required to conduct them. Recent advances in the development of nonisotopic labels for probes hold great promise for the future use of probes.

Acknowledgment

We thank George J. Jackson, Food and Drug Administration, Washington, D.C., for his many thoughtful comments on this manuscript.

References

Ahmed, A. A., M. K. Moustafa, and E. H. Marth. 1983. Incidence of *Bacillus cereus* in milk and some milk products. *J. Food Prot.* **46**:126–128.

Altboum, A., I. Hertman, and S. Sarid. 1985. Penicillinase plasmid-linked genetic determinants for enterotoxin B and C_1 production in *Staphylococcus aureus. Infect. Immunol.* **47**:514–521.

Archer, D. L., and J. E. Kvenberg. 1985. Incidence and cost of foodborne diarrheal disease in the United States. *J. Food Prot.* **48**:887–894.

Archer, D. L., and F. E. Young. 1988. Contemporary issues: Diseases with a food vector. *Clin. Microbiol. Rev.* **1**:377–398.

Arens, M., and E. M. Swierkosz. 1989. Detection of rotavirus by hybridization with a nonradioactive synthetic DNA probe and comparison with commercial enzyme immunoassays and silver stained polyacrylamide gels. *J. Clin. Microbiol.* **27**:1277–1279.

Arrand, J. E. 1985. Preparation of nucleic acid probes. In *Nucleic Acid Hybridization,* ed. B. D. Hames and S. J. Higgins, pp. 17–45. Washington, DC: IRL Press.

Aulisio, C. C. G., T. M. Lanier, and M. A. Chappel. 1982. *Yersinia enterocolitica* O:13 associated with an outbreak in three southern states. *J. Food Prot.* **45**:1263.

Aulisio, C. C. G., J. T. Stanfield, S. D. Weagent, and W. E. Hill. 1983. Yersinosis associated with tofu consumption: Serological, biochemical and pathogenicity studies of *Yersinia enterocolitica* isolates. *J. Food Prot.* **46**:226–230.

Baig, B. H., I. K. Wachsmuth, G. K. Morris, and W. E. Hill. 1986. Probing of *Campylobacter jejuni* with DNA coding for *Escherichia coli* heat-labile enterotoxin. *J. Infect. Dis.* **154**:542.

Bayles, K. W., and J. J. Iandolo. 1989. Genetic and molecular analyses of the gene encoding staphylococcal enterotoxin. *J. Bacteriol.* **171:**4799–4806.

Bean, N. H., P. M. Griffin, J. S. Golding, and C. B. Ivey. 1990. Foodborne disease outbreaks, 5-year summary, 1983–1987. *Morbid. Mortal. Weekly Rep.* **39** (No. SS-1):15–57.

Bessesen, M. T., Q. Luo, H. A. Rotbart, M. J. Blaser, and R. T. Ellison, III. 1990. Detection of *Listeria monocytogenes* by using the polymerase chain reaction. *Appl. Environ. Microbiol.* **56:**2930–2932.

Betley, M. J., and J. J. Mekalanos. 1985. Staphylococcal enterotoxin A is encoded by phage. *Science* **229:**185–187.

Betley, M. J., and J. J. Mekalanos. 1988. Nucleotide sequence of the type A staphylococcal enterotoxin gene. *J. Bacteriol.* **170:**34–41.

Betley, M. J., S. Lofdahl, B. N. Kreiswirth, M. S. Bergdoll, and R. P. Novick. 1984. Staphylococcal enterotoxin A gene is associated with a variable genetic element. *Proc. Natl. Acad. Sci. (USA)* **81:**5179–5183.

Bialkowska-Hobrzanska, H. 1987. Detection of enterotoxigenic *Escherichia coli* by dot blot hybridization with biotinylated DNA probes. *J. Clin. Microbiol.* **25:**338–343.

Billing, E., and W. A. Cuthbert. 1958. "Bitty" cream. The occurrence and significance of *Bacillus cereus* spores in raw milk supplies. *J. Appl. Microbiol.* **21:**65–78.

Binz, T., H. Kurazono, M. R. Popoff, M. W. Eklund, G. Sakaguchi, S. Kozaki, K. Krieglstein, A. Henschen, D. M. Gill, and H. Niemann. 1990. Nucleotide sequence of the gene encoding *Clostridium botulinum* neurotoxin type D. *Nucleic Acids Res.* **18:**5556.

Black, R. E., K. H. Brown, S. Becker, A. R. M. Alim Abdul, and I. Huq. 1982. Longitudinal studies of infectious diseases and physical growth of children in rural Bangladesh. II. Incidence of diarrhea and association with known pathogens. *Am. J. Epidemiol.* **115:**315–324.

Blaser, M. J., D. J. Duncan, G. Warren, and W.-L. L. Wang. 1983. Experimental *Campylobacter jejuni* infection of adult mice. *Infect. Immunol.* **39:**908–916.

Bohach, G. A., and P. M. Schlievert. 1987. Nucleotide sequence of staphylococcal enterotoxin C₁ gene and relatedness to other pyrogenic toxins. *Mol. Gen. Genet.* **209:**15–20.

Boileau, C. R., H. M. d'Hauteville, and P. J. Sansonetti. 1984. DNA hybridization technique to detect *Shigella* species and enteroinvasive *Escherichia coli*. *J. Clin. Microbiol.* **20:**959–961.

Brenner, D. J., A. G. Steigewalt, D. P. Falcao, R. E. Weaver, and G. R. Fanning. 1976. Characterization of *Yersinia enterocolitica* and *Yersinia pseudotuberculosis* by deoxyribonucleic acid hybridization and by biochemical reactions. *Int. J. Syst. Bacteriol.* **26:**180–194.

Britten, R. J., and D. E., Kohne. 1968. Repeated sequences in DNA. *Science* **161:**529–540.

Brown, J. E., P. Echeverria, D. N. Taylor, J. Seriwatana, V. Vanapruks, U. Lexomboon, R. N. Neill, and J. W. Newland. 1989. Determination by DNA hybridization of Shiga-like-toxin-producing *Escherichia coli* in children with diarrhea in Thailand. *J. Clin. Microbiol.* **27:**291–294.

Bruce, C., W. Al-Nakib, M. Forsyth, G. Stanway, and J. W. Almond. 1989. Detection of enteroviruses using cDNA and synthetic oligonucleotide probes. *J. Virol. Methods* **25:**233–240.

Bryan, F. L. 1979. Infections and intoxication caused by other bacteria. In *Foodborne Infections and Intoxications,* 2d ed., ed. H. Riemann and F. L. Bryan, pp. 211–297. New York: Academic Press.

Caldwell, M. B., R. I. Walker, S. D. Stewart, and J. E. Rogers. 1983. Simple adult rabbit model for *Campylobacter jejuni* enteritis. *Infect. Immunol.* **42:**1176–1182.

Calva, E., J. Torres, M. Vazquez, V. Angeles, H. de la Vega, and G. M. Ruiz-Palacios. 1989. *Campylobacter jejuni* chromosomal sequences that hybridize to *Vibrio cholerae* and *Escherichia coli* LT enterotoxin genes. *Gene* **75:**243–251.

Caruthers, M. H. 1985. Gene synthesis machines: DNA chemistry and its uses. *Science* **230:**281–285.

Cebula, T. A., and W. H. Koch. 1990. Analysis of spontaneous and psoralen-induced *Salmonella typhimurium hisG46* revertants by oligodeoxyribonucleotide colony hybridization: Use of psoralens to cross-link probes to target sequences. *Mutat. Res.* **229:**79–87.

Celum, C. L., R. E. Chaisson, G. W. Rutherford, J. L. Barnhart, and D. F. Echenberg. 1987. Incidence of salmonellosis in patients with AIDS. *J. Infect. Dis.* **156:**998–1002.

Centers for Disease Control. 1990. Community outbreaks of shigellosis—United States. *Morbid. Mortal. Weekly Rep.* **39:**509–519.

Chalker, R. B., and M. J. Blaser. 1988. A review of human salmonellosis. III. Magnitude of *Salmonella* infection in the United States. *Rev. Infect. Dis.* **10:**111–124.

Chan, S. W., S. Wilson, A.-Y. Hsu, W. King, D. N. Halbert, and J. Klinger. 1989. Model non-isotopic hybridization system for detection of foodborne bacteria: Preliminary results and future prospects. In *Biotechnology and Food Quality,* ed. S. D. Kung, D. D. Bills, and R. Quatron, pp. 219–237. Boston: Butterworths.

Chatterjee, N. K., M. Kaehler, and R. Deibel. 1988. Detection of enteroviruses using subgenomic probes of Coxsackie virus B4 by hybridization. *Diagn. Microbiol. Infect. Dis.* **11:**129–136.

Chevrier, D., D. Larzul, F. Megraud, and J.-L. Guesdon. 1989. Identification and classification of *Campylobacter* strains by using nonradioactive DNA probes. *J. Clin. Microbiol.* **27:**321–326.

Christiansson, A., A. S. Naidu, I. Nilsson, T. Wadstrom, and H.-E. Pettersson. 1989. Toxin production by *Bacillus cereus* dairy isolates in milk at low temperatures. *Appl. Environ. Microbiol.* **55:**2595–2600.

Cliver, D. O., R. D. Ellender, and M. D. Sobsey. 1984. Foodborne viruses. In *Compendium of Methods for the Microbiological Examination of Foods,* ed. M. L. Speck, pp. 508–541. Washington, DC: American Public Health Association.

Coghill, D., and H. S. Juffs. 1979. Incidence of psychrotrophic bacteria in pasteurized milk and cream products and effect of temperature on their growth. *Aust. J. Dairy Technol.* **34:**150–153.

Colwell, J. L., P. S. Grushoff-Kosyk, and A. W. Bernheimer. 1976. Purification of cereolysin and the electrophoretic separation of the active (reduced) and inactive (oxidized) forms of the purified toxin. *Infect. Immunol.* **14**:144–154.

Coolbaugh, J. C., and R. P. Williams. 1978. Production and characterization of two hemolysins of *Bacillus cereus*. *Can. J. Microbiol.* **24**:1289–1295.

Couch, J. L., M. T. Soltis, and M. J. Betley. 1988. Cloning and nucleotide sequence of the type E staphylococcal enterotoxin gene. *J. Bacteriol.* **170**:2954–2960.

Cover, T. L. and R. C. Aber. 1989. *Yersinia enterocolitica*. *New England J. Med.* **321**:16–24.

Cravioto, A., R. J. Gross, S. M. Scotland, and B. Rowe. 1979. An adhesive factor found in strains of *Escherichia coli* belonging to the traditional infantile entero-pathogenic serotypes. *Current Microbiol.* **3**:95–99.

Cukor, G., and N. R. Blacklow. 1984. Human viral gastroenteritis. *Microbiol. Rev.* **48**:157–179.

Curiale, M. S., M. J. Klatt, and M. A. Mozola. 1990. Colorimetric deoxyribonu-cleic acid hybridization assay for rapid screening of *Salmonella* in foods. *J. Assoc. Off. Anal. Chemists* **73**:248–256.

D'Aoust, J.-Y. 1989. Salmonella. In *Foodborne Bacterial Pathogens,* ed. M. P. Doyle, pp. 328–446. New York: Marcel Dekker.

Datta, A. R., B. A. Wentz, and W. E. Hill. 1987. Detection of hemolytic *Listeria monocytogenes* by using DNA colony hybridization. *Appl. Environ. Microbiol.* **53**:2256–2259.

Davies, F. L., and G. Wilkinson. 1973. *Bacillus cereus* in milk and dairy products. In *The Microbiological Safety of Food,* ed. B. C. Hobbs and J. H. B. Christian, pp. 57–68. New York: Academic Press.

DeBuono, B. A., J. Brondum, J. M. Kramer, R. J. Gilbert, and S. M. Opal. 1988. Plasmid, serotypic, and enterotoxin analysis of *Bacillus cereus* in an outbreak setting. *J. Clin. Microbiol.* **26**:1571–1574.

Dovey, S., and K. J. Towner. 1989. A biotinylated DNA probe to detect bacterial cells in artificially contaminated foodstuffs. *J. Appl. Bacteriol.* **66**:43–47.

Doyle, M. P. 1985. Food-borne pathogens of recent concern. In *Annual Review of Microbiology* vol. 39, ed. L. N. Ornston, A. Balows, and P. Baumann, pp. 25–41. Palo Alto, CA: Annual, Reviews.

Doyle, M. P. 1986. Detection and quantitation of foodborne pathogens and their toxins: Gram-negative bacterial pathogens. In *Foodborne Microorganisms and Their Toxins: Developing Methodologies,* ed M. D. Pierson and N. J. Stern, pp. 317–344. New York: Marcel Dekker.

Doyle, M. P., ed. 1989. *Foodborne Bacterial Pathogens.* New York: Marcel Dekker.

Echeverria, P., D. N. Taylor, J. Seriwatana, A. Charkaeomorakot, V. Khungvalert, T. Sakuldaipeara, and R. A. Smith. 1986. A comparative study of enterotoxin gene probes and tests for toxin production to detect ETEC. *J. Infect. Dis.* **153**:255–257.

Echeverria, P., D. N. Taylor, J. Seriwatana, J. E. Brown, and U. Lexomboon. 1989. Examination of colonies and stool blots for detection of enteropathogens by DNA hybridization with eight DNA probes. *J. Clin. Microbiol.* **27**:331–334.

Eiden, J. J., F. Firoozmand, S. Sato, S. L. Vonderfecht, F. Z. Yin, and R. H.

Yolken. 1989. Detection of group B rotavirus in fecal specimens by dot hybridization with a cloned cDNA probe. *J. Clin. Microbiol.* **27:**422–426.

Erlich, H. A., R. Gibbs, and H. H. Kazazian, Jr., eds. 1989. *Polymerase Chain Reaction.* Cold Spring Harbor, NY: Cold Spring Harbor Laboratory Press.

Feng, P. 1990. Detection of invasive *Yersinia* species using oligonucleotide probes. *J. Biol. Chem. Suppl. 14C,* p. 167 (Abstracts, UCLA Symposium on Molecular and Cellular Biology).

Ferreira, J. L., W. E. Hill, M. K. Hamdy, F. A. Zapatka, and S. G. McCay. 1986. Detection of enterotoxigenic *Escherichia coli* in foods by DNA colony hybridization. *J. Food Sci.* **51:**665–667.

Finch, M. H., and P. A. Blake. 1985. Foodborne outbreaks of campylobacteriosis: The United States experience, 1980–1982. *Am. J. Epidemiol.* **122:**262–268.

Fitts, R. 1985. Development of a DNA-DNA hybridization test for the presence of *Salmonella* in foods. *Food Technol.* **39:**95–102.

Fitts, R., M. Diamond, C. Hamilton, and M. Neri. 1983. DNA-DNA hybridization assay for detection of *Salmonella* in foods. *Appl. Environ. Microbiol.* **46:**1146–1151.

Flores, J., R. H. Purcell. I. Perez, R. G. Wyatt, E. Boeggeman, M. Sereno, L. White, R. M. Chanock, and A. L. Kapikian. 1983. A dot hybridization assay for detection of rotavirus. *Lancet* **1:**555–559.

Flowers, R. S. 1985. Comparison of rapid *Salmonella* screening methods and the conventional culture method. *Food Technol.* **39:**103–108.

Flowers, R. S., M. J. Klatt, M. A. Mozola, M. S. Curiale, D. A. Gabis, and J. H. Silliker. 1987. DNA hybridization assay for detection of *Salmonella* in foods: Collaborative study. *J. Assoc. Off. Anal. Chemists* **70:**521–529.

Frankel, G., L. Riley, J. A. Giron, J. Valmassoi, A. Friedmann, N. Strockbine, S. Falkow, and G. K. Schoolnik. 1990. Detection of *Shigella* in feces using DNA amplification. *J. Infect. Dis.* **161:**1252–1256.

Gatt, M. J., ed. 1984. Oligonucleotide synthesis: A practical approach. Washington, DC: IRL Press.

Gerba, P., B. Margolin, and E. Trumper. 1988. Enterovirus detection in water with gene probes. *Z. Gesamte Hyg.* **34:**518–519.

Gilmore, M. S., A. L. Cruz-Rodz, M. Leimeister-Wachter, J. Kreft, and W. Goebel. 1989. A *Bacillus cereus* cytolytic determinant, cereolysin AB, which comprises the phospholipase C and sphingomyelinase genes: Nucleotide sequence and genetic linkage. *J. Bacteriol.* **171:**744–753.

Gopo, J. M., R. Melis, E. Filipska, R. Meneveri, and J. Filipski. 1988. Development of a *Salmonella*-specific biotinylated DNA probe for rapid routine identification of *Salmonella. Mol. Cell. Probes* **2:**271–279.

Gouvea, V., R. I. Glass, P. Woods, K. Taniguchi, H. F. Clark, B. Forrester, and A.-Y. Fang. 1990. Polymerase chain reaction amplification and typing of rotavirus nucleic acid from stool specimens. *J. Clin. Microbiol.* **28:**276–282.

Grünstein, M., and D. S. Hogness. 1975. Colony hybridization: Method for the isolation of cloned DNAs that contain a specific gene. *Proc. Natl. Acad. Sci. (USA)* **85:**7652–7656.

Guatelli, J. C., K. M. Whitfield, D. Y. Kwoh, K. J. Barringer, D. D. Richman, and

T. R. Gingeras. 1990. Isothermal, *in vitro* amplification of nucleic acids by a multienzyme reaction modeled after retroviral replication. *Proc. Natl. Acad. Sci. (USA)* **87:**1874–1878.

Guth, B. E. C., C. L. Pickett, E. M. Twiddy, R. K. Holmes, T. A. T. Gomes, A. A. M. Lima, R. L. Guerrant, B. D. G. M. Franco, and L. R. Trabulsi. 1986. Production of type-II heat labile enterotoxin by *Escherichia coli* isolated from food and human feces. *Infect. Immunol.* **54:**587–589.

Hauschild, A. H. W, and F. L. Bryan. 1980. Estimate of cases of food- and water-borne illness in Canada and the United States. *J. Food Prot.* **43:**435–440.

Hauser, D., M. W. Eklund, J. Kurazono, T. Binz, J. Niemann, D. M. Gill, P. Boquet, and M. R. Popoff. 1990. Nucleotide sequence of *Clostridium botulinum* C1 neurotoxin. *Nucleic Acids Res.* **18:**4924.

Hill, W. E., and W. L. Payne. 1984. Genetic methods for the detection of microbial pathogens. Identification of enterotoxigenic *Escherichia coli* by DNA colony hybridization: Collaborative study. *J. Assoc. Off. Anal. Chemists* **67:**801–807.

Hill, W. E., W. L. Payne, and C. C. G. Aulisio. 1983. Detection and enumeration of virulent *Yersinia enterocolitica* in food by colony hybridization. *Appl. Environ. Microbiol.* **46:**636–641.

Hill, W. E., J. M. Madden, B. A. McCardell, D. B. Shah, J. A. Jagow, W. L. Payne, and B. K. Boutin. 1983. Foodborne enterotoxigenic *Escherichia coli:* Detection and enumeration by DNA colony hybridization. *Appl. Environ. Microbiol.* **45:**1324–1330.

Hill, W. E., W. L. Payne, G. Zon, and S. L. Moseley. 1985. Synthetic oligodeoxyribonucleotide probes for detecting heat-stable enterotoxin-producing *Escherichia coli* by DNA colony hybridization. *Appl. Environ. Microbiol.* **50:**1187–1191.

Hill, W. E., B. A. Wentz, W. L. Payne, J. A. Jagow, and G. Zon. 1986. DNA colony hybridization method using synthetic oligonucleotides to detect enterotoxigenic *Escherichia coli:* Collaborative study. *J. Assoc. Off. Anal. Chemists* **69:**531–536.

Hill, W. E., S. P. Keasler, E. L. Elliot, P. Feng, and K. A. Lampel. 1990. Detection of *Vibrio vulnificus* in oysters using the polymerase chain reaction. In *Abstracts of the Annual Meeting of the American Society for Microbiology,* p. 281, Washington, DC: American Society for Microbiology.

Hoge, C. W., O. Sethabutr, L. Bodhidatta, P. Echeverria, D. C. Robertson, and J. G. Morris, Jr. 1990. Use of a synthetic oligonucleotide probe to detect strains of non-serovar O1 *Vibrio cholerae* carrying the gene for heat-stable enterotoxin (NAG-ST). *J. Clin. Microbiol.* **28:**1473–1476.

Hsu, H.-Y., D. I. Sobell, S. W. Chan, J. McCarty, K. Parodos, D. J. Lane, and D. N. Halbert. 1989. A colorimetric DNA hybridization method for the detection of *Escherichia coli* in foods. In *Abstracts of the Annual Meeting of the American Society for Microbiology,* p. 322. Washington, DC: American Society for Microbiology.

Huang, I. Y., and M. S. Bergdoll. 1970. The primary structure of staphylococcal enterotoxin B. *J. Mol. Chem.* **243:**3518–3525.

Hyypia, T., P. Stalhandske, R. Vainionpaa, and U. Pettersson. 1984. Detection of enteroviruses by spot hybridization. *J. Clin. Microbiol.* **19:**436–438.

Ikezawa, H., M. Yamaanegi, R. Taguchi, T. Miyashita, and T. Ohyabu. 1976. Studies on phosphatidylinositol phosphodiesterase (phospholipase C type) of *Bacillus cereus*. *Biochim. Biophys. Acta* **450**:154–164.

Ikezawa, H., M. Mori, T. Ohyabu, and R. Taguchi. 1978. Studies on sphingomyelinase of *Bacillus cereus*. I. Purification and properties. *Biochim. Biophys. Acta* **528**:247–256.

Isberg, R. R., D. L. Voorhis, and S. Falkow. 1987. Identification of invasin: A protein that allows enteric bacteria to penetrate cultured mammalian cells. *Cell* **50**:769–778.

Izat, A. L., C. D. Driggers, M. Colberg, M. A. Reiber, and M. H. Adams. 1989. Comparison of the DNA probe to culture methods for the detection of *Salmonella* on poultry carcasses and processing waters. *J. Food Prot.* **52**:564–570.

Jackson, G. J., C. F. Langford, and D. L. Archer. 1991. Control of salmonellosis and similar foodborne infections. *Food Control* (Jan.), pp. 26–34.

Jacobs, K. A., R. Rudersdorf, S. D. Neill, J. P. Dougherty, E. L. Brown, and E. F. Fritsch. 1988. The thermal stability of oligonucleotide duplexes is sequence independent in tetraalkylammonium salt solutions: Application to identifying recombinant clones. *Nucleic Acids Res.* **16**:4637–4650.

Jagow, J. A., and W. E. Hill. 1986. Enumeration by DNA colony hybridization of virulent *Yersinia enterocolitica* colonies in artificially contaminated food. *Appl. Environ. Microbiol.* **51**:441–443.

Jagow, J. A., and W. E. Hill. 1988. Enumeration of virulent *Yersinia enterocolitica* colonies by DNA colony hybridization using alkaline treatment and paper filters. *Mol. Cell. Probes* **2**:189–195.

Jagow, J. A., and K. A. Lampel. 1989. Detecting enteroinvasive *Shigella* in food using a DNA probe. In *Abstracts of the Annual Meeting of the American Society for Microbiology,* p. 321. Washington, DC: American Society for Microbiology.

Jiang, X., M. K. Estes, and T. G. Metcalf. 1987. Detection of Hepatitis A virus by hybridization with single-stranded RNA probes. *Appl. Environ. Microbiol.* **53**:2487–2495.

Jiang, X., M. K. Estes, T. G. Metcalf, and J. L. Melnick. 1986. Detection of Hepatitis A virus in seeded estuarine samples by hybridization with cDNA probes. *Appl. Environ. Microbiol.* **52**:711–717.

Johnsson, K. M. 1984. *Bacillus cereus* foodborne illness—An update. *J. Food Prot.* **47**:145–153.

Jones, C. L., and S. A. Kahn. 1986. Nucleotide sequence of the enterotoxin B gene from *Staphylococcus aureus*. *J. Bacteriol.* **166**:29–33.

Kaper, J. B., S. L. Moseley, and S. Falkow. 1981. Molecular characterization of environmental and non-toxigenic strains of *Vibrio cholerae*. *Infect. Immunol.* **32**:661–667.

Kapperud, G., K. Dommarsnes, M. Skurnik, and E. Hornes. 1990. A synthetic oligonucleotide probe and a cloned polynucleotide probe based on the *yop*A gene for detection and enumeration of virulent *Yersinia enterocolitica*. *Appl. Environ. Microbiol.* **56**:17–23.

Kauffmann, F. 1966. *The Bacteriology of Enterobacteriaceae*. Baltimore, MD: Williams & Wilkins.

Kay, B. A., K. Wachsmuth, and P. Gemski. 1982. New virulence-associated plasmid in *Yersinia enterocolitica*. *J. Clin. Microbiol.* **15**:1161-1163.

Kaysner, C. A., C. Abeyta, Jr., M. M. Wekell, A. DePaola, Jr., R. F. Stott, and J. Leitch. 1987. Virulent strains of *Vibrio vulnificus* isolated from estuaries of the United States West Coast. *Appl. Environ. Microbiol.* **53**:1349-1351.

Keller, G. H., and M. M. Manak. 1989. *DNA Probes.* New York: Stockton Press.

Kirii, Y., H. Danbara, K. Komase, H. Arita, and M. Yoshikawa. 1987. Detection of enterotoxigenic *Escherichia coli* by colony hybridization with biotinylated enterotoxin probes. *J. Clin. Microbiol.* **25**:1962-1965.

Klipstein, F. A., and R. F. Engert. 1984. Purification of *Campylobacter jejuni* enterotoxin. *Lancet* **1**:1123-1124.

Korolik, V., P. J. Cole, and V. Krishnapillai. 1988. A specific DNA probe for the identification of *Campylobacter jejuni*. *J. Gen. Microbiol.* **134**:521-529.

Kramer, J. M., and R. J. Gilbert. 1989. *Bacillus cereus* and other *Bacillus* species. In *Foodborne Bacterial Pathogens,* ed M. P. Doyle, pp. 22-70. New York: Marcel Dekker.

Kramer, F. R. and P. M. Lizardi. 1989. Replicatable RNA reporters. *Nature* **339**:401-402.

Labbe, R. 1989. *Clostridium perfringens.* In *Foodborne Bacterial Pathogens,* ed. M. P. Doyle, pp. 191-234. New York: Marcel Dekker.

LaBrec, E. H., H. Schneider, T. J. Magnani, and S. B. Formal. 1964. Epithelial cell penetration as an essential step in the pathogenesis of bacillary dysentery. *J. Bacteriol.* **88**:1503-1518.

Lampel, K. A., J. A. Jagow, M. Trucksess, and W. E. Hill. 1990. Polymerase chain reaction for detection of invasive *Shigella flexneri* in food. *Appl. Environ. Microbiol.* **56**:1536-1540.

Larkin, E. P. 1981. Food contaminants—viruses. *J. Food Prot.* **44**:320-325.

Larkin, E. P. 1986. Detection, quantitation, and public health significance of foodborne viruses. In *Foodborne Microorganisms and Their Toxins: Developing Methodologies,* ed. M. D. Pierson and N. J. Stern, pp. 439-451. New York: Marcel Dekker.

Lee, E., B. McCardell, and P. Guerry. 1985. Characterization of a plasmid encoded enterotoxin in *Campylobacter jejuni.* In *Proceedings of the Third International Workshop on Campylobacter Infections,* ed. A. D. Pearson, M. B. Skirrow, H. Lior, and B. Rowe, p. 56. London: Public Health Laboratory Service.

Lee, C. H., S. L. Moseley, H. W. Moon, S. C. Whipp, C. L. Gyles, and M. So. 1983. Characterization of the gene encoding heat-stable toxin II and preliminary molecular epidemiological studies of enterotoxigenic *Escherichia coli* heat-stable toxin II producers. *Infect. Immunol.* **42**:264-268.

Levine, M. M., D. R. Nalin, R. B. Hornick, E. J. Bergquist, D. H. Waterman, C. R. Young, and S. Stoman. 1978. *Escherichia coli* strains that cause diarrhoea but do not produce heat-labile or heat-stable enterotoxins and are non-invasive. *Lancet* **1**:1119-1122.

Levine, M. M., J. G. Xu, J. B. Kaper, H. Lior, V. Prado, B. Tall, J. Nataro, J. Karch, and K. Wachsmuth. 1987. A DNA probe to identify enterohemorrhagic *Escherichia coli* of O157:H7 and other serotypes that cause hemorrhagic colitis and hemolytic uremic syndrome. *J. Infect. Dis.* **156**:175–182.

Maas, R. 1983. An improved colony hybridization method with significantly increased sensitivity for detection of single genes. *Plasmid* **10**:296–298.

Maas, R., R. M. Silva, T. A. T. Games, L. R. Trabulsi, and W. K. Maas. 1985. Detection of genes for heat-stable enterotoxin I in *Escherichia coli* strains isolated in Brazil. *Infect. Immunol.* **49**:46–51.

Mafu, A. A., R. Higgins, M. Nadeau, and G. Cousineau. 1989. The incidence of *Salmonella, Campylobacter,* and *Yersinia enterocolitica* in swine carcasses and the slaughterhouse environment. *J. Food Prot.* **52**:642–645.

Mallonee, D. H., B. A. Glatz, and P. A. Pattee. 1982. Chromosomal mapping of a gene affecting enterotoxin A production in *Staphylococcus aureus. Appl. Environ. Microbiol.* **43**:397–402.

Martinelli, R. A., E. Carroll III, and J. G. Donahue. 1990. A chemiluminescent DNA probe assay for *Campylobacter.* In *Abstracts of the Annual Meeting of the American Chemical Society,* section ANYL 54. Washington, DC: American Chemical Society.

Matthews, J. A., and L. J. Kricka. 1988. Analytical strategies for the use of DNA probes. *Anal. Biochem.* **169**:1–25.

McCarthy, B. J. 1967. Arrangement of base sequences in deoxyribonucleic acid. *Bacteriol. Rev.* **31**:215–229.

McManus, C., and J. M. Lanier. 1987. *Salmonella, Campylobacter jejuni* and *Yersinia enterocolitica* in raw milk. *J. Food Prot.* **50**:51–55.

Mehlman, I. J. 1984. *Shigella.* In *Bacteriological Analytical Manual,* 6th ed., pp. 9.01–9.05. Arlington, VA: Association of Official Analytical Chemists.

Meinkoth, J., and G. Wahl. 1984. Hybridization of nucleic acids immobilized on solid supports. *Anal. Biochem.* **138**:267–284.

Miliotis, M. D., J. E. Galen, J. B. Kaper, and J. Glenn Morris, Jr. 1989. Development and testing of a synthetic oligonucleotide probe for the detection of pathogenic *Yersinia* strains. *J. Clin. Microbiol.* **27**:1667–1670.

Miliotis, M. D., J. G. Morris, S. Cianciosi, A. C. Wright, P. K. Wood, and R. M. Robins-Browne. 1990. Identification of a conjunctivitis-associated gene locus from the virulence plasmid of *Yersinia enterocolitica. Infect. Immunol.* **58**:2470–2477.

Miller, J. K., and W. M. Barnes. 1986. Colony probing as an alternative to standard sequencing as a means of direct analysis of chromosomal DNA to determine the spectrum of single-base changes in regions of known sequence. *Proc. Natl. Acad. Sci. (USA)* **83**:1026–1030.

Miller, V. L., J. J. Farmer, III, W. E. Hill, and S. Falkow. 1989. The *ail* locus is found uniquely in *Yersinia enterocolitica* serotypes commonly associated with disease. *Infect. Immunol.* **57**:121–131.

Miyada, C. G., A. B. Studencki, and R. B. Wallace. 1987. Synthetic oligonucleotides for the identification and isolation of specific gene sequences. In *Synthesis and Applications of DNA and RNA,* ed. S. A. Narang, pp. 207–227. New York: Academic Press.

Miyamoto, Y., T. Kato, Y. Obara, S. Akiyama, K. Takizawa, and S. Yamai. 1969.

In vitro hemolytic characteristic of *Vibrio parahaemolyticus:* Its close correlation with human pathogenicity. *J. Bacteriol.* **100**:1147–1149.

Morotomi, M., S. Hoshina, P. Green, H. C. Neu, P. LoGerfo, I. Watanabe, M. Mutai, and I. B. Weinstein. 1989. Oligonucleotide probe for detection and identification of *Campylobacter pylori. J. Clin. Microbiol.* **27**:2652–2655.

Morris, J. G., Jr., A. C. Wright, D. M. Roberts, P. K. Wood, L. M. Simpson, and J. D. Oliver. 1987. Identification of environmental *Vibrio vulnificus* isolates with a DNA probe for the cytotoxin-hemolysin gene. *Appl. Environ. Microbiol.* **53**:193–195.

Moseley, S. L., P. Echeverria, J. Seriwatana, C. Tirapat, W. Chaicumpa, T. Sakulaipaera, and S. Falkow. 1982. Identification of enterotoxigenic *Escherichia coli* by colony hybridization using three enterotoxin gene probes. *J. Infect. Dis.* **145**:863–869.

Moseley, S. L., and S. Falkow. 1980. Nucleotide sequence homology between the heat-labile enterotoxin gene of *Escherichia coli* and *Vibrio cholerae* deoxyribonucleic acid. *J. Bacteriol.* **144**:444–446.

Moseley, S. L., I. Huq, A.R.M.A. Alim, M. So, M. Samadpour-Motalebi, and S. Falkow. 1980. Detection of enterotoxigenic *Escherichia coli* by DNA colony hybridization. *J. Infect. Dis.* **142**:892–898.

Nataro, J. P., M. M. Baldini, J. B. Kaper, R. E. Black, N. Bravo, and M. M. Levine. 1985. Detection of an adhesive factor of enteropathogenic *Escherichia coli* with a DNA probe. *J. Infect. Dis.* **152**:560–565.

Neill, R. J., G. R. Fanning, F. Delahoz, R. Wolff, and P. Gemski. 1990. Oligonucleotide probes for detection and differentiation of *Staphylococcus aureus* strains containing genes for enterotoxins A, B, and C and toxic shock syndrome toxin I. *J. Clin. Microbiol.* **28**:1514–1518.

Newland, J. W., and R. J. Neill. 1988. DNA probes for Shiga-like toxins I and II and for toxin-converting bacteriophages. *J. Clin. Microbiol.* **26**:958–987.

Nishibuchi, M., and J. B. Kaper. 1985. Nucleotide sequence of the thermostable direct hemolysin of *Vibrio parahaemolyticus. J. Bacteriol.* **162**:558–564.

Nishibuchi, M., M. Ishibashi, Y. Takeda, and J. B. Kaper. 1985. Detection of the thermostable direct hemolysin gene and related DNA sequences in *Vibrio parahaemolyticus* and other *Vibrio* species by the DNA colony hybridization test. *Infect. Immunol.* **49**:481–486.

Nishibuchi, M., W. E. Hill, G. Zon, W. L. Payne, and J. B. Kaper. 1986. Synthetic oligodeoxyribonucleotide probes to detect Kanagawa phenomenon-positive *Vibrio parahaemolyticus. J. Clin. Microbiol.* **23**:1091–1095.

Notermans, S., K. J. Heuvelman, and K. Wernars. 1988. Synthetic enterotoxin B DNA probes for detection of enterotoxigenic *Staphylococcus aureus* strains. *Appl. Environ. Microbiol.* **54**:531–533.

O'Brien, A., and R. K. Holmes. 1987. Shiga and Shiga-like toxins. *Microbiol. Rev.* **51**:206–220.

Olive, D. M., M. Johny, and S. K. Sethi. 1990. Use of an alkaline phosphatase-labeled synthetic oligonucleotide probe for detection of *Campylobacter jejuni* and *Campylobacter coli. J. Clin. Microbiol.* **28**:1565–1569.

Olsvik, Ø., K. Wachsmuth, G. Morris, and J. C. Feeley. 1984. Genetic probing of

Campylobacter jejuni for cholera toxin and *Escherichia coli* heat-labile entero-toxin. *Lancet* 1:449.

Oosterom J., S. Notermans, H. Karman, and G. B. Engels. 1983. Origin and preva-lence of *Campylobacter jejuni* in poultry processing. *J. Food Prot.* 46:339–344.

Otnaess, A.-B., C. Little, K. Slettin, R. Wallin, W. Johnson, R. Flengsrud, and H. Prydz. 1977. Some characteristics of phospholipase C from *Bacillus cereus*. *Eur. J. Biochem.* 79:459–468.

Overcast, W. W., and K. Atmaram. 1974. The role of *Bacillus cereus* in sweet cur-dling of fluid milk. *J. Milk Food Technol.* 37:233–236.

Panda, C. S., L. W. Riley, S. N. Kunmari, K. K. Khanna, and K. Prakash. 1990. Comparison of alkaline phosphatase-conjugated oligonucleotide DNA probe with the Sereny test for identification of *Shigella* strains. *J. Clin. Microbiol.* 28:2122–2124.

Petitjean, J., M. Quibriac, F. Freymuth, F. Fuchs, N. Laconche, M. Aymard, and H. Kopecka. 1990. Specific detection of enteroviruses in clinical samples by mo-lecular hybridization using Poliovirus subgenomic riboprobes. *J. Clin. Micro-biol.* 28:307–311.

Picken, R. N., Z. Wang, and H. L. Yang. 1987. Molecular cloning of a species-specific DNA probe for *Campylobacter jejuni. Mol. Cell. Probes* 1:245–259.

Portnoy, D. A., S. A. Moseley, and S. Falkow. 1981. Characterization of plasmid and plasmid-associated determinants of *Yersinia enterocolitica* pathogenesis. *In-fect. Immunol.* 31:775–782.

Portnoy, D. A., H. Wolf-Watz, I. Bolin, A. B. Beeder, and S. Falkow. 1984. Char-acterization of common virulence plasmids in *Yersinia* species and their role in the expression of outer membrane proteins. *Infect. Immunol.* 43:108–114.

Ranelli, D. M., C. L. Jones, M. R. Johns, G. J. Mussey, and S. A. Kahan. 1985. Molecular cloning of staphylococcal enterotoxin B gene in *Escherichia coli* and *Staphylococcus aureus. Proc. Natl. Acad. Sci. (USA)* 82:5850–5854.

Rashtchian, A., and M. S. Curiale. 1989. DNA probes assays for detection of *Cam-pylobacter* and *Salmonella*. In *Nucleic Acid and Monoclonal Antibody Probes,* ed. B. Swaminathan and G. Prakash, pp. 221–239, New York: Marcel Dekker.

Rashtchian, A., M. A. Abbott, and M. Shaffer. 1987. Cloning and characterization of genes coding for ribosomal RNA in *Campylobacter jejuni. Current Microbiol.* 14:311–317.

Renz, M., and C. Kurz. 1985. A colormetric method for DNA hybridization. *Nu-cleic Acids Res.* 12:3435–3444.

Riley, L. K., and C. J. Caffrey. 1990. Identification of enterotoxigenic *Escherichia coli* by colony hybridization with nonradioactive digoxigenin-labeled DNA probes. *J. Clin. Microbiol.* 28:1465–1468.

Roberts, T. 1989. Human illness costs of foodborne bacteria. *Am. J. Agric. Econ.* 71:468–474.

Robins-Browne, R., M. D. Miliotis, S. Cianciosi, V. L. Miller, S. Falkow, and J. G. Morris, Jr. 1989. Evaluation of DNA colony hybridization and other tech-niques for detection of virulence in *Yersinia* species. *J. Clin. Microbiol.* 27:644–650.

Robinson, D. A. 1981. Infective dose of *Campylobacter jejuni*. *Brit. Med. J.* **282**:1584.

Romaniuk, P. J., and T. J. Trust. 1989. Rapid identification of *Campylobacter* species using oligonucleotide probes to 16S ribosomal RNA. *Mol. Cell. Probes* **3**:133–142.

Romick, T. L., J. A. Lindsay, and F. F. Busta. 1989. Evaluation of a visual DNA probe for enterotoxigenic *E. coli* detection in foods and wastewater by colony hybridization. *J. Food Prot.* **52**:466–470.

Rotbart, H. A., M. J. Levin, and L. P. Villarreal. 1984. Use of subgenomic poliovirus DNA hybridization probes to detect the major subgroups of enteroviruses. *J. Clin. Microbiol.* **20**:1105–1108.

Ruiz-Palacios, G. M., E. Escamilla, and N. Torres. 1981. Experimental *Campylobacter* diarrhea in chickens. *Infect. Immun.* **34**:250–255.

Ruiz-Palacios, G. M., J. Torres, N. I. Torres, E. Escamilla, B. R. Ruiz-Palacios, and J. Tamayo. 1983. Cholera-like enterotoxin produced by *Campylobacter jejuni*. *Lancet* **2**:250–251.

Ryster, E. T., and E. H. Marth. 1989. "New" food-borne pathogens of public health significance. *Perspect. Pract.* **7**:948–954.

Saito, M. 1990. Production of enterotoxin by *Clostridium perfringens* derived from humans, animals, foods, and the natural environment in Japan. *J. Food Prot.* **53**:115–118.

Saiki, R. K., D. H. Gelfand, S. Stoffel, S. J. Scharf, R. Higuchi, G. T. Horn, K. Mullis, and H. A. Erlich. 1988. Primer-directed enzymatic amplification of DNA with a thermostable DNA polymerase. *Science* **239**:487–491.

St. Louis, M. E., D. L. Morse, M. E. Potter, T. M. DeMelfi, J. J. Guzewich, R. V. Tauxe, and P. A. Blake. 1988. The emergence of grade A eggs as a major source of *Salmonella enteritidis* infections. New implications for the control of salmonellosis. *J. Am. Med. Assoc.* **259**:2103–2107.

Sall, B. S., M. Lombardo, B. Sheridan, and G. H. Parsons. 1988. Performance of a DNA probe to detect *Salmonella typhi*. *J. Clin. Microbiol.* **22**:600–605.

Samadpour, M., J. Liston, J. E. Ongerth, and P. I. Tarr. 1990. Evaluation of DNA probes for detection of Shiga-like-toxin-producing *Escherichia coli* in food and calf fecal samples. *Appl. Environ. Microbiol.* **56**:1212–1215.

Sansonetti, P. J., D. J. Kopecko, and S. B. Formal. 1981. *Shigella sonnei* plasmids: Evidence that a large plasmid is necessary for virulence. *Infect. Immunol.* **34**:75–83.

Schiemann, D. A. 1989. *Yersinia enterocolitica* and *Yersinia pseudotuberculosis*. In *Foodborne Bacterial Pathogens*, ed. M. P. Doyle, pp. 601–672. New York: Marcel Dekker.

Schmidt, J. J., and L. Spero. 1983. The complete amino acid sequence of staphylococcal enterotoxin C_1. *J. Biol. Chem.* **258**:6300–6305.

Scully, M. L., C. Chapis, and A. P. Kuritza. 1989. Identification of *Salmonella* species with a synthetic oligonucleotide probe. In *Abstracts of the Annual Meeting of the American Society for Microbiology*, p. 112. Washington, DC: American Society for Microbiology.

Seriwatana, J., P. Echeverria, D. N. Taylor, T. Sakuldaipeara, S. Changchawalit, and O. Chivoratanind. 1987. Identification of enterotoxigenic *Escherichia coli* with synthetic alkaline phosphatase-conjugated oligonucleotide DNA probes. *J. Clin. Microbiol.* **25**:1438–1441.

Shafer, W. M., and J. J. Iandolo. 1978. Staphylococcal enterotoxin A: A chromosomal gene product. *Appl. Environ. Microbiol.* **36**:389–391.

Skirrow, M. B. 1977. *Campylobacter* enteritis: A "new" disease. *Brit. Med. J.* **2**:9–11.

Small, P. C., and S. Falkow. 1986. Development of a DNA probe for the virulence plasmid of *Shigella* spp. and enteroinvasive *Escherichia coli*. In *Microbiology*, D. Schlessinger, ed. pp. 121–124. Washington, DC: American Society for Microbiology.

Smith, H. R., S. M. Scotland, H. Chart, and B. Rowe. 1987. Vero cytotoxin production and presence of VT genes in strains of *Escherichia coli* and *Shigella*. *FEMS Microbiol. Lett.* **42**:173–177.

Smith, J. L. 1987. *Shigella* as a foodborne pathogen. *J. Food Prot.* **50**:788–801.

Sommerfelt, H., H. M. S. Grewal, and M. K. Bhan. 1990. Simplified and accurate nonradioactive polynucleotide gene probe assay for identification of enterotoxigenic *Escherichia coli*. *J. Clin. Microbiol.* **28**:49–54.

Sowers, E., N. Strockbine, D. Cameron, J. Green, T. A. Cebula, and W. L. Payne. 1989. Comparison of oligonucleotide and natural fragment probes for detection SLT-I and SLT-II produced by *Escherichia coli*. In *Abstracts of the Annual Meeting of the American Society for Microbiology*, p. 112. Washington, DC: American Society for Microbiology.

Spira, W. M., and J. M. Goepfert. 1975. Biological characteristics of an enterotoxin produced by *Bacillus cereus*. *Can. J. Microbiol.* **21**:1236–1246.

Stolzenbach, F., L. A. P. Phillips, Y. Y. Yang, R. K. Enns, and M. S. You. 1988. Highly specific DNA probes for the detection and differentiation of *Campylobacter* species. In *Abstracts of the Annual Meeting of the American Society for Microbiology*, p. 282. Washington, DC: American Society for Microbiology.

Tacket, C. O., F. Brenner, and P. A. Blake. 1984. Clinical features and an epidemiological study of *Vibrio vulnificus* infections. *J. Infect. Dis.* **149**:558–561.

Taylor, D. N., P. Echeverria, O. Sethabutr, C. Pitarangsi, U. Leksomboon, N. R. Blacklow, B. Rowe, R. Gross, and J. Cross. 1988. Clinical and microbiologic features of *Shigella* and enteroinvasive *Escherichia coli* infections detected by DNA hybridization. *J. Clin. Microbiol.* **26**:1362–1366.

Tenover, F. C. 1988. Diagnostic deoxyribonucleic acid probes for infectious diseases. *Clin. Microbiol. Rev.* **1**:82–101.

Tenover, F. C., L. Carlson, S. Barbagallo, and I. Nachamkin. 1990. DNA probe culture confirmation assay for identification of thermophilic *Campylobacter* species. *J. Clin. Microbiol.* **28**:1284–1287.

Thompson, D. E., J. K. Brehm, J. D. Oultram, T.-J. Swinfield, C. C. Shone, T. Atkinson, J. Melling, and N. P. Minton. 1990. The complete amino acid sequence of the *Clostridium botulinum* type A neurotoxin, deduced by nucleotide sequence analysis of the encoding gene. *Eur. J. Biochem.* **189**:73–81.

Todd, E. C. D. 1989. Preliminary estimates of costs of foodborne disease in the United States. *J. Food Prot.* **52**:595–601.

Tompkins, L. S., N. Troup, A. Labigne-Roussel, and M. L. Cohen. 1986. Cloned, random chromosomal sequences as probes to identify *Salmonella* species. *J. Infect. Dis.* **154**:156–162.

Trucksess, M. W., K. M. Williams, B. A. Wentz, and W. E. Hill. 1988. A synthetic probe for detection of enterotoxigenic *Staphylococcus aureus* in foods. In *Abstracts of the Annual Meeting of the American Chemical Society*, section AGFD 69. Washington, DC: American Chemical Society.

Tsubokura, M., K. Otsuki, K. Sato, M. Tanaka, T. Hongo, H. Fukushima, T. Maruyama, and M. Inoue. 1989. Special feature of distribution of *Yersinia pseudotuberculosis* in Japan. *J. Clin. Microbiol.* **27**:790–791.

Turnbull, P. C. B., J. M. Kramer, K. Jorgensen, R. J. Gilbert, and J. Melling. 1979. Properties and production characteristics of vomiting, diarrheal, and necrotizing toxins of *Bacillus cereus*. *Am. J. Clin. Nutr.* **32**:219–228.

Van Damme-Jongsten, M., J. Rodhouse, R. J. Gilbert, and S. Notermans. 1990. Synthetic DNA probes for detection of enterotoxigenic *Clostridium perfringens* strains isolated from outbreaks of food poisoning. *J. Clin. Microbiol.* **28**:131–133.

Venkatesan, M., J. M. Buysee, E. Vandendries, and D. J. Kopecko. 1988. Development and testing of invasion-associated DNA probes for detection of *Shigella* spp. and enteroinvasive *Escherichia coli*. *J. Clin. Microbiol.* **26**:261–266.

Venkatesan, M., J. M. Buysee, E. Vandendries, and D. J. Kopecko. 1989. Use of *Shigella flexneri ipaC* and *ipaH* gene sequences for the general identification of *Shigella* spp. and enteroinvasive *Escherichia coli*. *J. Clin. Microbiol.* **27**:2687–2691.

Wallace, R. B., J. Shaffer, R. F. Murphy, J. Bonner, T. Hirose, and K. Itakura. 1979. Hybridization of synthetic oligodeoxyribonucleotides to ØX174 DNA. The effect of single base pair mismatch. *Nucleic Acids Res.* **6**:3543–3557.

Wernars, K., and S. Notermans. 1990. Gene probes for detection of food-borne pathogens. In *Gene Probes for Bacteria*, ed. A. J. de Macario and E. C. De Macario, pp. 353–388. New York: Academic Press.

White, P. B. 1926. *Great Britain Medical Research Council, Special Report No. 103*. London: Her Majesty's Stationery Office.

Wilson, K. H., R. B. Blitchington, and R. C. Greene. 1990. Amplification of bacterial 16S ribosomal DNA with polymerase chain reaction. *J. Clin. Microbiol.* **28**:1942–1946.

Wong, H.-C., M.-H. Chang, and J.-Y. Fan. 1988. Incidence and characterization of *Bacillus cereus* isolates contaminating dairy products. *Appl. Environ. Microbiol.* **54**:699–702.

Wood, W. I., J. Gitschier, L. A. Lasky, and R. M. Lawn. 1985. Base composition-independent hybridization in tetramethylammonium chloride: A method for oligonucleotide screening of highly complex gene libraries. *Proc. Natl. Acad. Sci. (USA)* **82**:1585–1588.

Wood, P. K., J. G. Morris, P. L. C. Small, O. Sethabutr, M. R. F. Toledo, L. Trabulsi, and J. B. Kaper. 1986. Comparison of DNA probes and the Sereny test for identification of invasive *Shigella* and *Escherichia coli* strains. *J. Clin. Microbiol.* **24**:498–500.

Wright, A. C., J. G. Morris, Jr., D. R. Maneval, Jr., K. Richardson, and J. B. Kaper. 1985. Cloning of the cytotoxin-hemolysin gene of *Vibrio vulnificus*. *Infect. Immunol.* **50**:922–924.

Yamamoto, K., A. C. Wright, J. B. Kaper, and J. G. Morris, Jr. 1990. Cytolysin gene of *Vibrio vulnificus:* Sequence and relationship to *V. cholerae* El Tor hemolysin. *Infect. Immunol.* **58**:2706–2709.

Zheng, B. J., W. P. Lam, Y. K. Yan, S. K. F. Lo, M. L. Lung, and M. H. Ng. 1989. Direct identification of serotypes of natural human rotavirus isolates by hybridization using cDNA probe derived from segment 9 of the rotavirus genome. *J. Clin. Microbiol.* **27**:552–557.

8

Rapid Methods for the Detection of *Listeria*

Jerrie Gavalchin, Katrina Landy, and Carl A. Batt

The presence of *Listeria monocytogenes* in milk and other food products represents an ever increasing food safety problem. Infants and immuno-compromised persons are at particular risk (Gellin and Broome 1989). The techniques currently available for the detection and enumeration of *L. monocytogenes* are not rapid enough to assure the safety of the product prior to its consumption. These techniques are only useful in diagnosing the probable etiological agent following a suspected food illness outbreak. Therefore, the need to develop rapid and accurate methods for the detection of *Listeria* are of paramount importance. The obvious market for these tests has prompted an intense effort in both commercial and academic laboratories to establish such methods. This chapter is an attempt to objectively review all of the literature on the subject published to date without the normal biases associated with such an effort. An excellent review of a number of microbiological, epidemiological and detection aspects of *Listeria* has been compiled by Miller, Smith, and Somkuti (1990).

ECOLOGY AND PATHOGENESIS OF *LISTERIA*

Listeria is a gram-positive bacillus, and while most species within the genus are nonpathogenic to humans, *Listeria monocytogenes* has the potential to cause listeriosis. It is primarily a livestock pathogen, but it has been seen sporadically in humans for many years. The earliest cases of listeriosis can be traced back to the late 1800s, and its initial identification as the causative agent in human disease to the 1920s. Foodborne outbreaks in Germany originating principally from the consumption of raw milk have been documented as far back as the 1930s–1950s. It was not until recent foodborne outbreaks in the United States, Canada, and Europe that interest in *Listeria* has intensified.

Listeriosis is primarily a disease of immunocompromised individuals, transplant patients, pregnant women and their fetuses, and the elderly. The disease is characterized by septicemia, meningitis, and abortion, and can result in death. The most severe consequence of infection by *L. monocytogenes* in an average person is most often flu-like symptoms. It has been established from the outbreaks that the mortality rate of this disease in "immunesusceptible" individuals can be as high as 31%.

The genus *Listeria* includes a number of species, only some of which are pathogenic. It is increasingly clear that *L. monocytogenes* is the predominant (and perhaps only) pathogenic species found in foods (Cox et al. 1989). *L. monocytogenes* is genetically diverse, as demonstrated by allelic variation in a number of metabolic enzymes (Piffaretti et al. 1989). The different serotypes represent a population of two distinct phylogenetic divisions, making the identification of a common antigen potentially difficult. The selective detection of *L. monocytogenes* is of great importance, as the ecology of *Listeria* is not well understood and the detection of *Listeria* spp. other than *L. monocytogenes,* may not be an indicator of a food's safety. Some evidence supporting this concept has recently been reported, based upon ecological screening for a number of *Listeria* spp. (Breer and Schopfer 1988; Cox et al. 1989).

The ubiquitous nature of *Listeria* is documented by a number of recent surveys demonstrating its presence in a variety of foods. The foodborne outbreaks in the last 10 years have primarily involved dairy-based products, although coleslaw is another single food of notoriety. A number of factors direct attention toward dairy products, including the propensity for *Listeria* to grow at refrigerated temperatures, the frequent occurrence of *Listeria* in raw milk, and the consumption of milk-based products by a susceptible population. Ideally, a detection method sufficiently rapid to detect *Listeria* in dairy products would ensure a level of safety prior to its release for sale. Although this is the ideal, the most likely use would be in a hazard analysis critical control points (HACCP) approach where critical elements and ingredients in the process were monitored as an indication of a problem. Whether the development of such a methodology is technically feasible is only part of the issue, the other being the ability to eradicate *Listeria* from the food supply.

MICROBIOLOGICAL METHODS
FOR DETECTION OF *LISTERIA*

The study of *Listeria* with respect to its detection and any attempts to target it as the etiological agent in suspected foodborne outbreaks present several problems. Its detection is difficult because it is found in very low numbers within food. Since multiplication is inevitable in favorable conditions, even the presence of a single bacterial cell could be a potential problem, necessitating a very sensitive methodology. Other microflora in the food, including other *Listeria* species, make detection of the pathogenic *L. monocytogenes* a complex, time-consuming, and labor-intensive process. This organism may have an incubation period of up to 30 days, making it difficult to establish a link between the onset of the disease and the contaminated food. *L. monocytogenes* is found ubiquitously, making establishment of the disease source difficult.

Current methodology for the detection and enumeration of *Listeria* involves enrichment in selective media, which may include incubation at refrigeration temperatures (Klinger 1988). Selective media that allow only *Listeria* to grow on them have also been developed. The wide range of media that may be used for detection of *Listeria* is daunting. A number of comparative studies have been reported; however, no single methodology appears to be so vastly superior as to be adopted universally (Brackett et al. 1990). Another issue of emerging importance is the recovery of injured *Listeria* cells. Sublethal thermal processing in addition to other intrinsic and/ or extrinsic factors can injure *Listeria*. The phenomenon of injury is not new, but the potential significance of injured *Listeria* in foods will need to be considered in the formulation of enrichment/recovery media.

The first step in most protocols for identification of *Listeria* begins with an enrichment phase to enhance the number of *Listeria* as compared to other microflora. Enrichment broths include FDA, UVM, Doyle and Schoeni, and Fraser (Buchanan 1990). Most of these contain a combination of antibiotics, including nalidixic acid, acriflavin, or polymyxin B to suppress the growth of other microflora. Selective plating media, including modified McBride, LPM, and modified Vogel Johnson agars contain similar antibiotics. In addition selective plating media may contain compounds such as esculin, whose hydrolysis is indicative of *Listeria*. The most recent FDA recommendations include the use of Oxford medium as a substitute for modified McBride. Initial identification of *Listeria* is usually followed by a series of biochemical tests to confirm its identification and to establish the species. The time required for final confirmation may exceed several weeks (where cold enrichment is employed), depending upon the initial population, the type of food, and the contaminating microflora. This protracted analysis time precludes a real-time analysis of a number of perishable yet suspect products, such as ready-to-eat foods and dairy products.

RAPID METHODS FOR THE DETECTION OF *LISTERIA* SPP.

A number of efforts to detect *Listeria* using either antibodies (monoclonal or polyclonal) or nucleic acid probes have been reported (Tables 8–1 and

TABLE 8-1   Summary of Nucleic Acid Probes for the Detection of *Listeria*

Specificity	Target	Format	Reference
Listeria sp.	16s rRNA	Solution	King et al. 1990
L. monocytogenes (*L. seeligeri*)	*iap*	Colony Southern Dig-colony	Datta et al. 1988 Kohler et al. 1990 Kim et al. 1991
L. monocytogenes	*hly*A	Southern Colony	Chenevert et al. 1991 Datta, Wentz, and Russell, 1990
L. monocytogenes L. ivanovii	DTH	Colony	Notermans et al. 1989
L. monocytogenes	*hly*A	PCR	Bessesen et al. 1990

8-2). There are several philosophical differences between the selection of either nucleic acid probes or antibodies for the detection of pathogenic microorganisms and their products. In fact neither approach is an absolute indicator of the safety of a food product. One can imagine a scenario where either test renders a positive result, when in actuality the organism is not viable and therefore not a threat. Since all of these "rapid" detection systems require some preenrichment, growth of the targeted organism is mandated, and the problem of "dead" bacteria is not an issue.

Nucleic-Acid-based Probes

Nucleic acid probes have in the past few years become a viable tool for detecting a number of virus, bacteria, and other microorganisms in either food, clinical, or environmental samples. The target in these detection sys-

TABLE 8-2   Summary of Monoclonal Antibodies Reactive with *Listeria*

Specificity	Format	Reference
Listeria sp.	ELISA	Butman et al. 1988
Listeria sp.	ELISA	Farber and Speirs 1987
L. monocytogenes Serotype 1/2, 3, or 4, 4b	Fluorescence Microscopy	McLauchlin and Pini 1989
L. monocytogenes L. welshimeri L. innocua	Western Immuno-dot-blot	Siragusa and Johnson 1990
L. monocytogenes	ELISA Immunomagnetic	Ziegler and Orlin 1984 Skjerve, Rorvik, and Olsvik 1990
L. monocytogenes (All serotypes)	ELISA	This chapter

tems can be any one of a number of nucleic acids, including ribosomal RNA, mitochondrial DNA, plasmid DNA, or chromosomal DNA. The key criterion for the selection of any target nucleic acid is that it uniquely defines the organism in question with little or no probability of existing in another microorganism that might be found in the same ecological niche. There is obviously no way of ensuring that a targeted nucleic acid sequence will be found only in the microorganism for which the detection system is being developed. This is especially true where the sequence is cryptic and chosen simply on the basis of its uniqueness within a selected test population. In cases where a specific toxin gene sequence is selected, there is an assumption that it will not be widely distributed in nature.

The use of 16s rRNA as a distinct signature for a bacterium has become the method of choice when no other obvious nucleic acid sequence uniquely defines the desired target (Woese 1987). Databases of 16s rRNA sequences covering a wide diversity of microorganisms are available and can be used to search for regions that are characteristic of the targeted microorganism. A DNA probe based upon the sequence for the 16s rRNA that detects all *Listeria* spp. (Klinger 1988; Klinger et al. 1988) has been developed by GeneTrak Inc. (Framingham, Massachusetts). Although the exact sequence is proprietary, it is clearly derived from one of the variable regions of the 16s rRNA. Initial detection systems using this 16s rRNA sequence involved the use of ^{32}P-labeled probes, which limited its utility, especially in the food-plant environment. Subsequently, a novel solution hybridization assay has been formatted where final quantification is accomplished using an enzymatic marker (King et al. 1990). Briefly, the 16s rRNA is released by alkaline lysis from cells grown in an enrichment broth. Then a capture tag consisting of the complementary sequence to a unique region of the 16s rRNA and a poly A tail is allowed to hybridize to the target 16s rRNA. This hybrid is then removed from solution via the poly A tail using a poly T sequence immobilized on a polystyrene solid support. The detection is accomplished using an antibody coupled to horseradish peroxidase directed against a fluorescein marker covalently linked to the detector probe. The detector probe recognizes sequences in the 16s rRNA as spatially distinct from the region recognized by the capture probe. Therefore the oxidation of a substrate (tetramethyl benzidine) in the presence of hydrogen peroxide by horseradish peroxidase can be used as an indication of the presence of *Listeria*. A potential refinement of this approach would involve the use of a 16s rRNA probe specific for *L. monocytogenes*. Some efforts toward characterizing such a probe have been reported, although they are far too preliminary to warrant further comment (Pandian and Emond 1990).

A probe derived from a putative delayed hypersensitivity factor (DTH) isolated from *L. monocytogenes* 1/2a hybridizes to all serogroups of *L.*

monocytogenes in addition to *L. ivanovii,* but not to any other *Listeria* spp. tested (Notermans et al. 1989). The exact nature of the gene coding for this delayed hypersensitivity factor has not yet been reported, therefore its role in the pathogenicity of *L. monocytogenes* cannot be determined. It does, however, appear to be an effective tool for detecting *Listeria,* although its species specificity is not absolute for *L. monocytogenes.* For example, the DTH gene appears to be absent from *L. monocytogenes* serogroup 4a, yet present in *L. ivanovii.* At this time its use as a nucleic acid probe has been limited to colony hybridization assays, and the entire 1.1-kb DTH-containing fragment labeled with ^{32}P served as the probe.

A sequence from what was first believed to be a putative *L. monocytogenes* α-hemolysin gene (Flamm 1986) has been reported to be specific for *L. monocytogenes* (Datta, Wentz, and Hill 1987; Datta et al. 1988). Subsequent analysis did not establish it as a hemolysin, and it was termed a *major secreted protein* (msp) (Flamm, Hinrichs, and Thomashow 1989). Despite its nebulous character, this sequence has proved useful in developing a nucleic-acid-based detection system for *Listeria.* Initially, a colony hybridization protocol was used where suspect colonies were transferred to nitrocellulose filters and probed with this ^{32}P-labeled fragment. Good specificity. was shown toward *Listeria* spp. that were β-hemolytic (CAMP-positive). Subsequent refinements of the approach have included the evaluation of four synthetic 20-bp oligonucleotide probes in lieu of the entire 500-bp fragment. Two probes that were tested against a range of *Listeria* spp. hybridized to all isolates of *L. monocytogenes* and one weakly hemolytic isolate of *L. seeligeri* (Datta et al. 1988). Recently, the origin of this probe has been clarified by the reported cloning and sequencing of an invasion-associated protein (*iap*) by Kohler et al. (1990). The *iap* gene contains the sequences originally used by Datta et al. (1988) as oligonucleotide probes for detection of *L. monocytogenes.* Notable in the amino acid sequence of the *iap* protein are 19 copies of a Thr-Asn dipeptide repeat whose function is not known. A 400-bp *Hind*III fragment, as compared to a larger 1.6-kb *Dde*I fragment from the *iap* gene, appears to be useful for detection of *L. monocytogenes,* as demonstrated by southern hybridization (Kohler et al. 1990).

The pathogenicity of *L. monocytogenes* is probably dependent upon a number of factors, including the production of at least one hemolysin. Transposon mutagenesis (Tn916) disrupting the coding sequence for listeriolysin O renders *L. monocytogenes* avirulent in mice. The gene coding for listeriolysin O has been cloned (Mengaud et al. 1988; Leimeister-Wachter and Chakroborty 1989; Datta, Wentz, and Russell 1990) and sequenced (Mengaud et al. 1988). The 1617-bp open reading frame codes for a 58-kD (kilodalton) protein with the expected amino terminal signal sequence to direct its secretion. Interestingly, when the listeriolysin O gene is introduced into *Bacillus subtilis,* it confers the ability to grow in macrophage-like cells in culture (Bielecki et al. 1990).

The listeriolysin O gene appears to be unique to *L. monocytogenes,* and is an obvious target for developing a detection system. It does, however, share some amino acid homology with other hemolysins, including strepto-lysin O and pneumolysin. The listeriolysin O gene has been used as a probe in southern hybridization analysis of DNA purified from a number of *Listeria* sp (Chenevert et al. 1989). A 610-bp fragment internal to the region coding for listeriolysin O appears to hybridize only with hemolytic strains of *L. monocytogenes.* However, under nonstringent conditions, a probe derived from sequences on the 3' of the listeriolysin O gene was able to hybridize with hemolytic strains of *L. ivanovii* and *L. seeligeri.* The implica-tion is that there is some nucleotide sequence conservation between the hemolysin genes in *Listeria,* although the exact extent of homology will be determined only after sequence analysis of these other hemolysins is com-pleted. In an extension of the use of listeriolysin O as a probe, Datta, Wentz, and Russell (1990) used two synthetic oligonucleotide probes for detection of *L. monocytogenes.* Using colony hybridization techniques sim-ilar to those described (Datta et al. 1988), good specificity for only *L. monocytogenes* was obtained. As discussed in more detail later, it is likely that a listeriolysin probe will be amenable to a number of assay formats applica-ble to analyzing food samples.

The sensitivity of a nucleic-acid-based detection system is a function of a number of parameters, including the number of copies of the target within a single cell. The use of 16s rRNA has the obvious advantage of there being more than 100 copies of it within a cell. This assay, therefore, will be much more sensitive than an assay based upon a single-copy target. Polymerase chain reaction (PCR) involves the enzymatic amplification of a targeted nucleic acid sequence using a thermostable DNA polymerase and flanking oligonucleotide primers that uniquely define the target. The most com-monly used DNA polymerase is from *Thermus aquaticus,* and is termed *taq* polymerase. Since the amplification is exponential, by using a cyclic series of denaturing, annealing, and extension steps, the target can be amplified in excess of a millionfold with respect to the other sequences within the cell. The power of PCR prompted its obvious application for the detection of *L. monocytogenes.* Bessesen et al. (1990) designed primers based upon the listeriolysin O sequences as originally reported by Mengaud et al. (1988). Total DNA isolated from very dilute cultures was amplified approximately 30 cycles. The amplified products separated by agarose gel electrophoresis, and then visualized by ethidium bromide staining, were of the 606-bp size as expected based upon the intervening distance between the primers. As expected, this approach greatly increases the sensitivity of the assay, and approximately 10^4 cells could be detected. The challenge now, however, is to apply this technique to routine food samples to determine the efficacy of PCR to amplify DNA extracted from cells in this medium. It is unclear whether the reaction conditions, and especially the *taq* polymerase, can work effectively in the presence of the food constituents. Furthermore, given the extreme sensitivity of PCR, routine problems of cross contamination of samples through, for example, aerosols (an event common in even the most sophisticated molecular biology laboratories) must be addressed.

Antibody-based Detection Systems

The use of immunological assays for the detection of bacteria is not new and has been a part of, for example, the classic methodology for *Salmonella*. The traditional method (although no longer in widespread use) for detection of *Salmonella* involves a series of enrichment and selective plating media followed by a fluorescent polyclonal antibody assay for final confirmation of the organism.

The limitations of polyclonal antiserum are obvious; they tend to cross-react with a number of other bacteria, obviating their use for the primary identification of a genus. They have been used, however, for serological analysis of isolates, proving effective in establishing the epidemiological relationship between suspected outbreaks of foodborne illnesses. Perhaps one of the earliest demonstrations of immunological detection of *Listeria* was reported by Eveland (1963). *L. monocytogenes* was detected in the spinal fluid from a patient with meningitis using rabbit polyclonal antibodies raised against heat-treated *L. monocytogenes* (serotypes 1, 2, 3, 3b, 4a, and 4b) conjugated directly to fluorescein isothiocyanate.

The prospects of using immunological reagents as a rapid method for detection of microorganisms improved dramatically with the advent of hybridoma technology (Kohler, Howe, and Milstein 1976). Monoclonal antibodies are produced by hybridomas, which are the result of fusing an antibody secreting lymphocyte to a plasmacytoma cell. These hybridomas can then be grown in culture and the antibody harvested from the medium. Alternatively, the hybridoma cells can be injected into an appropriately primed mouse and allowed to establish a tumor. This ascites tumor is a highly productive source of monoclonal antibodies.

The first monoclonal antibody against *Listeria* antibodies were raised by immunization with semipurified flagella extracts of *Listeria* spp. (Farber and Speirs 1987; Table 8–2). These react with all *Listeria* spp. except *L. grayi, L. murrayi,* and *L. denitrificans*. They do not react with any other gram-positive bacteria tested, including *Staphylococcus aureus* and *Strepto-*

coccus faecalis. An assay using these antibodies has been formatted where the bacteria are spotted onto a nitrocellulose filter and then detected with the monoclonal antibody in conjunction with a secondary peroxidase-coupled antibody.

A monoclonal antibody has been characterized that reacts with a heat-stable antigen from *Listeria* (Butman et al. 1988; Mattingly et al. 1988). The antibody was raised by immunizing mice with a heat-treated lysate of *L. monocytogenes* and subsequently fusing the splenocytes to a mouse myeloma fusion partner. Hybridomas were screened by a direct binding assay to a heat-treated lysate from a culture of *L. monocytogenes*. Although the exact nature of the antigen is not known, it has been commercialized by Organon-Teknika and is available. It requires (as do all rapid methods developed to date) a preenrichment step, after which the culture is collected and the extract produced by heating. The detection of this *Listeria* antigen is accomplished using an enzyme-linked immunosorbent assay (ELISA) format with two different monoclonal antibodies to first capture the antigen and to then subsequently detect the trapped antigen. The two different monoclonal antibodies recognize different epitopes on the antigen, thereby avoiding competition. The monoclonal antibody used for detection is directly conjugated to horseradish peroxidase, and tetramethylbenzidine used as the chromogenic substrate. The total time involved in the actual assay is approximately 2 hours. The heat-stable nature of this antigen (which varies in size from 30 kDa to 38 kDa) is potentially problematic for samples heavily contaminated with *Listeria,* which are then thermally processed. In this situation, a false-positive reaction is possible, although given the current need for enrichment, only viable *Listeria* will be detected. Some efforts to document the utility of this assay have been reported (Beumer and Brinkman 1989).

Hybridomas that produced monoclonal antibodies that reacted with *L. monocytogenes* were isolated by Ziegler and Orlin (1984) by immunization of mice with both live and heat-killed *L. monocytogenes.* Only a limited number of *Listeria* spp., including *L. grayi, L. denitrificans,* and *L. murrayi,* were tested using a radioimmunoassay and shown not to cross-react with some of the antibodies isolated. McLauchlin et al. (1988) characterized two monoclonal antibodies, CL17, which recognized *L. monocytogenes* serotypes 1/2, 3, and CL2, which reacted with serotypes 4b, 4 (not 4b), and *L. innocua* serovar 6a. These could be used to detect *L. monocytogenes* in soft cheese by direct fluorescence microscopy, although some samples that were known to contain *L. monocytogenes* based upon standard microbiological tests were negative by their analysis (McLauchlin and Pini 1989).

As mentioned previously, the selective detection of *L. monocytogenes* is of great importance, and for this reason we have isolated a monoclonal antibody that will specifically recognize *L. monocytogenes* (Gavalchin et al. 1988, 1991). We have characterized several murein monoclonals produced by fusing spleen cells isolated from Balb/c mice immunized with live *L. monocytogenes* to NS-1 plasmacytoma cells. The immunogen was live *L. monocytogenes* Scott A cells that were injected directly into the spleen of the mice (Spitz et al. 1984). The choice of using live cells (as opposed to heat- or formalin-killed) and the route of immunization was to provide the most direct presentation of an unaltered (or minimally processed) antigen to the spleen. Hybridomas were screened by direct ELISA assay (Epstein and Lunney 1985), and of the 150 hybridomas tested, three reacted most strongly with *L. monocytogenes* Scott A. Although Mab 20-10-2, Mab 36-6-12, and Mab 59-9-16 reacted to some extent with *L. innocua* and *L. ivanovii* in the direct binding assay, greater specificity for *L. monocytogenes* was seen in an indirect ELISA assay. These antibodies were used to trap *L. monocytogenes,* which were then detected by a rabbit anti-*L. monocytogenes* polyclonal antiserum.

There are a number of parameters to be optimized in order to increase the sensitivity and specificity of these monoclonal antibodies. In our indirect ELISA, we are using rabbit anti-*L. monocytogenes* polyclonal antiserum and a secondary goat–antirabbit antibody for detection. Greater specificity will probably be achieved using a cocktail of monoclonal antibodies to detect *L. monocytogenes* trapped using a monoclonal antibody. Furthermore, the monoclonal antibodies used for detection of the trapped antigen could be either conjugated directly to the reporter molecule or biotinylated to utilize the sensitive biotin-strepavidin amplification system. In addition to these improvements, the incubation conditions, buffers (ionic strength, pH), and other assay parameters could be manipulated to improve the assay. Currently, we are also pursuing monoclonal antibodies raised against other antigens, which on the basis of two-dimensional western blotting, appear unique to *L. monocytogenes.*

We do not know the nature of the antigen(s) that are recognized by Mab 20-10-2, Mab 36-6-12, and Mab 59-9-16. Since all of the assays are performed using intact microorganisms, we suspect that the antigen(s) are exposed on the surface of the cell. The monoclonal antibodies fail to react when tested by SDS-PAGE western blotting of *L. monocyogenes* extracts. This suggests that the antigens may be a conformational determinant that is lost during denaturation or a composite of two distinct macromolecules that are dissociated during boiling in SDS. Their heat lability may prove to be advantageous for detecting *L. monocytogenes* in processed foods, since the antigen (as well as the organism) would presumably not survive processing. Another important issue is the physiological state of the cell upon enrichment. Where cell injury is suspected, some preenrichment might be advisable in order for the cell to recover and grow in the selective enrichment medium.

One difficulty we have encountered with the monoclonal antibodies that we have isolated is that they are all of the IgM class and are therefore relatively unstable. An interesting observation we have made is that the medium used for culture enrichment can apparently modulate the amount of target antigen produced. For example, TSY-grown *L. monocytogenes* cells show a much higher reactivity with Mab 36-6-12 as compared to TSB-grown cells (Fig. 8–1). This is not surprising, since the culture conditions can modulate the cell surface characteristics of bacteria. For example, Peel, Donachie, and Shaw (1988) reported that the appearance of flagella in *L. monocytogenes* is temperature dependent and maximally produced at 20°C. In the case of all of the monoclonal antibody-based detection systems where the nature of the antigen is not clear, attention should be paid to the enrichment medium and its affect on antigen production.

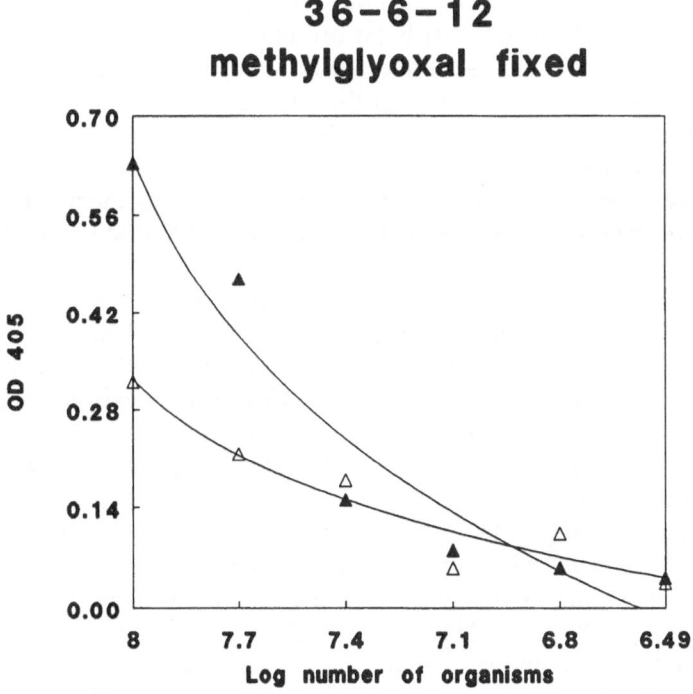

FIGURE 8-1.   Direct binding ELISA detection of *L. monocytogenes* grown in either TSB (△) or TSY (▲) media. Cells were fixed using methylglyoxal and MAB 36-6-12 as a primary antibody. Bound antibody was detected using a goat–antimouse horseradish peroxidase conjugate and tetramethylbenzidine as a chromogen.

In a similar manner, Siragusa and Johnson (1990) have attempted to isolate a monoclonal antibody specific for *L. monocytogenes*. These antibodies were the result of immunizations with heat-treated *L. monocytogenes* in a manner similar to previous work of Butman et al. (1988). Both immuno-dot-blot of heat-treated whole cells and western analysis of SDS-PAGE-separated cell extracts were used to demonstrate reactivity. Unfortunately, their monoclonal antibody reacted not only with *L. monocytogenes* but also with *L. welshmerii, L. innocua,* and possibly others. The presence of the common 18.5-kD antigenic epitope recognized by this monoclonal antibody is unfortunate, since the utility of this monoclonal antibody is limited.

THE REAL WORLD: LIMITATIONS AND PROBABLE IMPROVEMENTS

Obviously, the long-term goals of any program to develop a rapid method for the detection of any microorganism include that it be fast, simple, sensitive, accurate, and for commercial purposes, inexpensive. At least some of these desired performance attributes are mutually exclusive; for example, as an assay is made more sensitive, the accuracy as defined by the number of false positives increases. Most of the attention in the use of either monoclonal antibodies or nucleic acid probes for the identification of *Listeria* (or, in fact, other microorganisms) is focused on the reporter molecules and associated detection instrumentation. The advances in chemiluminescent-based reporters that have sensitivities in excess of 100-fold greater than existing enzymatic-based systems will be applicable to the *Listeria* problem. Another avenue for improvement is the specific concentration of the target antigen. To this end a number of approaches are feasible, one of which, employing an immunomagnetic bead separation, was demonstrated by Skjerve, Rørvik, and Olsvik (1990). Capitalizing on an existing technology where immunoglobulins can be fixed to a magnetic bead (2–3 μm) that can then be easily separated from the bulk solution, monoclonal antibodies (which albeit have not been extensively characterized) against *L. monocytogenes* were used for the selective concentration of the target cells. Approximately 200 cells per milliliter could be detected using this approach.

All of the rapid assays to date require a prior enrichment step, in certain cases of up to 48 hours. Therefore any claims that an assay can be completed within, for example, 4 hours, are not entirely truthful. There continues to be extensive effort to further improve the formulation of media for recovery of *Listeria* from foods, and this should prove beneficial as a prelude to any rapid detection method. Another area of concern is the significance of injured *Listeria* cells in a given food product and their potential for recovery either during enrichment or long-term in the food during storage. As mentioned previously, antibodies or nucleic acid probes can in theory detect not only injured cells but dead cells as well. If in the future a rapid assay is developed that can directly detect a microorganism in a food without any prior enrichment, the issue of the significance of an injured population will need to be addressed.

It is difficult to make an accurate comparison between the relative advantages and disadvantages of the two test kits currently in commercial use from GeneTrak and Organon-Teknika. They are both priced in the $5-to-$10 per-test range, and the recommended enrichment steps prior to the actual assay are similar. Both kits appear to be usable by a person with only minimal training in microbiology and still yield qualitative results. It should be noted that neither test kit has received Association of Official Analytical Chemists (AOAC) approval. It is interesting to note that both companies market competing products for the detection of *Salmonella*. Since neither kit has taken a substantial command of the market share, the clear advantage of one approach over the other may be a subjective decision by the end user.

Finally, the question of detection systems that are specific for *L. monocytogenes* raises some interesting issues. The most obvious argument for an *L. monocytogenes*-specific test is based upon observations that *L. monocytogenes* is most frequently if not exclusively associated with human listeriosis. In an ideal world, a rapid test that detects *Listeria* spp. that are pathogenic in humans would be the goal. Until we have elucidated the factors mediating pathogenicity of *L. monocytogenes,* such a goal is not feasible.

Acknowledgements

The support of the Wisconsin Milk Marketing Board, International Life Sciences Institute (ILSI), and the Northeast Dairy Foods Research Center is greatly appreciated. The authors thank Mary Lou Tortorello, Renae Malek, Ruth Knight, Margaret Landers, and Sharon Best for their technical support, and Atin Datta for sharing his unpublished results. The authors also thank Barbara Russell for her help in the preparation of this manuscript.

References

Bessesen, M. T., Q. Luo, H. A. Rotbart, M. J. Blaser, and R. T. Ellison, III. 1990. Detection of *Listeria monocytogenes* by using the polymerase chain reaction. *Appl. Environ. Microbiol.* **56**:2930-2932.

Beumer, R. R., and E. Brinkman. 1989. Detection of *Listeria* spp. with a monoclonal antibody-based enzyme-linked immunosorbent assay (ELISA). *Food Microbiol.* **6**:171-177.

Bielecki, J., P. Youngman, P. Connelly, and D. A. Portnoy. 1990. *Bacillus subtilis* expressing a haemolysin gene from *Listeria monocytogenes* can grow in mammalian cells. *Nature* 345:175–176.

Brackett, R. E., L. R. Beuchat, D. A., Golden, and P. K. Cassiday. 1990. Assessment of the ability of plating methods to accurately detect *Listeria* in foods. In *Foodborne Listeriosis,* ed. A. L. Miller, J. L. Smith, and G. A. Somkuti, pp. 97–103. New York: Elsevier.

Breer, C., and K. Schopfer. 1988. *Listeria* and foods. *Lancet* (Oct. 29), p. 1022.

Buchanan, R. L. 1990. Advances in cultural methods for the detection of *Listeria monocytogenes*. In *Foodborne Listeriosis,* ed. A. L. Miller, J. L. Smith, and G. A. Somkuti, pp. 85–95. New York: Elsevier.

Butman, B. T., M. C. Plank, R. J. Durham, and J. A. Mattingly. 1988. Monoclonal antibodies which identify a genus-specific *Listeria* antigen. *Appl. Environ. Microbiol.* 54:1564–1569.

Chenevert, J., J. Mengaud, E. Gormley, and P. Cossart. 1989. A DNA probe specific for *L. monocytogenes* in the genus *Listeria. Intl. J. Food Microbiol.* 8:317–319.

Cox, L. J., T. Kleiss, J. L. Cordier, C. Cordellana, T. Konkel, C. Pedrazzini, R. Beumer, and A. Siebenga. 1989. *Listeria spp.* in food processing, non-food and domestic environments. *Food Microbiol.* 6:49–61.

Datta, A. R., B. A. Wentz, and W. E. Hill. 1987. Detection of hemolytic *Listeria monocytogenes* by using DNA colony hybridization. *Appl. Environ. Microbiol.* 53:2256–2259.

Datta, A. R., B. A. Wentz, D. Shook, and M. W. Trucksess, 1988. Synthetic oligodeoxribonucleotide probes for detection of *Listeria monocytogenes. Appl. Environ. Micro.* 54:2933–2937.

Datta, A. R., B. A. Wentz, and J. Russell, 1990. Cloning of the listerolysin O gene and development of species gene probes for *Listeria monocytogenes. Appl. Environ. Micro.* 56:3874–3877.

Epstein, S. L. and J. K. Lunney, 1985. A cell surface ELISA in the mouse using only poly-L-lysine as cell fixative. *J. Immunol. Methods* 76:63–71.

Eveland, W. C. 1963. Demonstration of *Listeria monocytogenes* in direct examination of spinal fluid by fluorescent antibody technique. *J. Bacteriol.* 85:1448–1450.

Farber, J. M. and J. I. Speirs, 1987. Monoclonal antibodies directed against the flagellar antigens of *Listeria* species and their potential in EIA-based methods. *J. Food Prot.* 50:479.

Flamm, R. K. 1986. Molecular genetics of *Listeria monocytogenes:* cloning of a hemolysin gene, demonstration of conjugation and detection of native plasmids. Ph.D. thesis. Washington State University, Pullman, WA.

Flamm, R. K., D. J. Hinrichs, and M. F. Thomashow. 1989. Cloning of a gene encoding a major secreted polypeptide of *Listeria monocytogenes* and its potential use as a species-specific probe. *Appl. Environ. Microbiol.* 55:2251–2256.

Gavalchin, J., M. L. Tortorello, M. Landers, and C. A. Batt. 1988. Development of a monoclonal antibody-based detection system for *Listeria monocytogenes,* Abstract 173. In *Proceedings, 48th Annual Institute of Food Technologists Meeting,* p. 123. New Orleans: IFT.

Gavalchin, J., M. L. Tortorello, R. Malek, M. Landers, and C. A. Batt. 1991. Isolation of monoclonal antibodies that react preferentially with *Listeria monocytogenes. Food Microbiol.* **8**:325–330.

Gellin, B. G., and C. V. Broome. 1989. Listeriosis. *J. Am. Med. Assoc.* **261**:1313–1319.

Kim, C., B. Swaminathan, P. K. Cassaday, L. W. Mayer, and B. P. Holloway. 1991. Rapid confirmation of *Listeria monocytogenes* isolated from foods by a colony blot assay using a digoxigenin-labeled synthetic oligonucleotide probe. *Appl. Environ. Microbiol.* **57**:1609–1614.

King, W., S. M. Raposa, J. E. Warshaw, A. R. Johnson, D. Lane, J. D. Klinger, and D. N. Halbert. 1990. A colorimetric assay for the detection of *Listeria* using nucleic acid probes. In *Foodborne Listeriosis,* ed. A. L. Miller, J. L. Smith, and G. A. Somkuti, pp. 117–124. New York: Elsevier.

Klinger, J. D. 1988. Isolation of *Listeria:* A review of procedures and future prospects. *Infection* **16**:S98–S105.

Klinger, J. D., A. Johnson, D. Croan, P. Flynn, K. Whippie, M. Kimball, J. Lawrie, and M. Curiale. 1988. Comparative studies of nucleic acid hybridization assay for *Listeria* in foods. *J. Assoc. Off. Anal. Chemists* **7**:669–673.

Kohler, G., S. S. Howe, and C. Milstein, 1976. Fusion between immunoglobulin-secreting and nonsecreting myeloma cell lines. *Eur. J. Immunol.* **6**:292–295.

Kohler, S., M. Leimester-Wachter, T. Chakraborty, F. Lottspeich, and W. Goebel. 1990. The gene coding for protein p60 of *Listeria monocytogenes* and its use as a species specific probe for *Listeria monocytogenes. Infect. Immunol.* **58**:1943–1950.

Leimeister-Wachter, M., and T. Chakroborty. 1989. Detection of listeriolysin, the thiol-dependent hemolysin in *Listeria monocytogenes, Listeria ivanovii, Listeria seeligeri. Infect. Immunol.* **57**:2350–2357.

Mattingly, J. A., B. T. Butman, M. C. Plank, R. J. Durham, and B. J. Robison. 1988. Rapid monoclonal antibody-based enzyme-linked immunosorbent assay for detection of *Listeria* in food products. *J. Assoc. Off. Anal. Chemists* **71**:679.

McLauchlin, J., and P. N. Pini. 1989. The rapid demonstration and presumptive identification of *Listeria monocytogenes* in food using monoclonal antibodies in a direct immunofluorescence test (DIFT). *Lett. Appl. Microbiol.* **8**:25–27.

McLauchlin, J., A. Black, H. T. Green, J. Q. Nash, and A. G. Taylor. 1988. Monoclonal antibodies show *Listeria monocytogenes* in necropsy tissue samples. *J. Clin. Pathol.* **41**:983–988.

Mengaud, J., M. Vicente, J. Chenevert, J. M. Pereira, C. Geoffrey, B. Gicquel-Sanzey, F. Baquero, J. Perez-Diaz, and P. Cossart. 1988. Expression in *Escherichia coli* and sequence analysis of the Listeriolysin O determinant of *Listeria monocytogenes. Infect. Immunol.* **56**:766–772.

Miller, A. L., J. L. Smith, and G. A. Somkuti, eds. 1990. *Foodborne Listeriosis.* New York: Elsevier.

Notermans, S., T. Chakraborty, M. Leimeister-Wachter, J. Dufrenne, K. J. Heuvelman, H. Maas, W. Jansen, K. Wernars, and P. Guinee. 1989. Specific gene probe for detection of biotyped and serotyped *Listeria* strains. *Appl. Environ. Microbiol.* **55**:902–906.

Pandian, S., and E. Emond. 1990. Identification of the rRNA genes of *Listeria monocytogenes*. In *Foodborne Listeriosis,* ed. A. L. Miller, J. L. Smith, and G. A. Somkuti, pp. 131–135. New York: Elsevier.

Peel, M., W. Donachie, and A. Shaw. 1988. Temperature-dependent expression of flagella of *Listeria monocytogenes* studied by electron microscopy, SDS-PAGE and western blotting. *J. Gen. Microbiol.* **134**:2171–2178.

Piffaretti, J-C., H. Kressebuch, M. Aeschbacher, J. Bille, E. Bannerman, J. M. Musser, R. K. Selander, and J. Rocourt. 1989. Genetic characterization of clones of the bacterium *Listeria monocytogenes* causing epidemic disease. *Proc. Natl. Acad. Sci. (USA)* **86**:3818–3822.

Siragusa, G. R., and M. G. Johnson. 1990. Monoclonal antibody specific for *Listeria monocytogenes, Listeria innocua,* and *Listeria welshimeri. Appl. Environ. Microbiol.* **56**:1897–1904.

Skjerve, E., L. M. Rørvik, and O. Olsvik. 1990. Detection of *Listeria monocytogenes* in foods by immunomagnetic separation. *Appl. Environ. Microbiol.* **56**:3478–3481.

Spitz, M., L. Spitz, R. Thorpe, and E. Egui. 1984. Intrasplenic primary immunization for the production of monoclonal antibodies. *J. Immunol. Methods* **70**:39–43.

Woese, C. R. 1987. Bacterial evolution. *Microbiol. Rev.* **51**:221–271.

Ziegler, H. K., and C. A. Orlin. 1984. Analysis of *Listeria monocytogenes* antigens with monoclonal antibodies. *Clin. Invest. Med.* **7**:239–242.

9

Molecular Strategies for Reducing Aflatoxin Levels in Crops before Harvest

Thomas E. Cleveland and Deepak Bhatnagar

Aflatoxins, produced by the filamentous fungi *Aspergillus flavus* and *A. parasiticus,* are potent toxins and carcinogens that contaminate food and feed worldwide (Jelinek, Pohland, and Wood 1989). Aflatoxin B1 is the most toxic of the aflatoxin family of compounds (see Fig. 9–1) and is usually the predominant compound; however, other "naturally produced" aflatoxins (B2, G1, and G2) are also toxic and contaminate food and feed. The aflatoxin family of compounds can form "adducts" with animal and human DNA (Hsu et al. 1991; Bressac et al. 1991) and can cause primary liver cancer in certain animal systems (Groopman and Sabbioni 1991; Wogan 1991). In human systems, the carcinogenicity of aflatoxins is less clear (Campbell et al. 1990), but certain associations between aflatoxin intake by human populations and primary liver cancer have been reported (Groopman and Sabbioni 1991; Yeh et al. 1989). With possible implications to human and animal health worldwide, intense efforts are underway to remove these compounds from animal and human food chains.

Two basic approaches are being used in the attempt to reduce or remove the threat of aflatoxin contamination of foods and feeds: the prevention of preharvest contamination of crops with aflatoxin (Lisker and Lillehoj 1991), and the detoxification of aflatoxin (Park et al. 1988) in already-contaminated seed after harvest. Researchers around the world are using approaches involving conventional methods, such as improved cultural practices, breeding for resistance, and the use of pesticides, as well as contemporary methods in biotechnology in attempts to remove aflatoxin from the food supply.

Detoxification (Park et al. 1988) and absorptive (Phillips et al. 1991) removal of aflatoxins from already-contaminated foods and feeds are two promising methods currently under extensive investigation; these methods

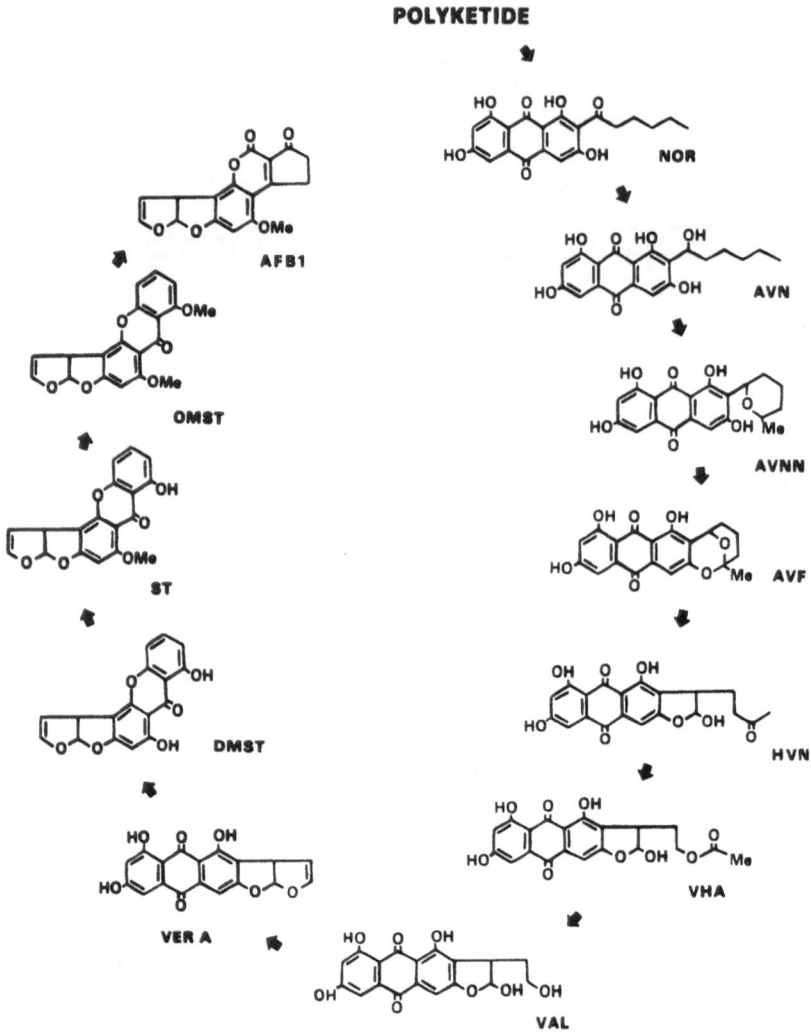

FIGURE 9-1.   The generally accepted scheme of known precursors in the biosynthesis of aflatoxin B_1. NOR, norsolorinic acid; AVN, averantin; AVNN, averufanin; AVF, averufin; HVN, 1′-hydroxyversicolorone; VHA, versiconal hemiacetal acetate; VAL, versiconal; Ver A, versicolorin A; DMST, demethylsterigmatocystin; ST, sterigmatocystin; OMST, *O*-methylsterigmatocystin; AFB_1, aflatoxin B_1.

could be important control measures for immediate application, since less progress has been made in developing practical methods to prevent preharvest aflatoxin contamination of foods and feeds. However, prevention of aflatoxin contamination before harvest is probably the best long-term approach; this strategy would obviate the need to detoxify large quantities of aflatoxin-contaminated seed material and also avoid the uncertainties of gaining approval from regulatory agencies for the use of detoxified seed for animal or human food. Therefore, this chapter focuses on the longer term strategy namely, the preharvest control of aflatoxin contamination.

PREHARVEST CONTROL OF AFLATOXIN CONTAMINATION

Conventional Technology for Preharvest Control of Aflatoxin

Conventional methods that are currently being utilized successfully to control many plant pathogens in the field have not been effective in field control of aflatoxin-producing fungi that infect peanut, corn, and cottonseed, the three major crops susceptible to preharvest aflatoxin contamination. Plant breeding programs have provided germ plasm that shows less susceptibility to aflatoxin contamination (Gorman and Kang 1991; Scott et al. 1991; Widstrom 1987), but none of the genetic material is "immune" to infection by *A. flavus* or *A. parasiticus* and/or aflatoxin contamination. However, the results of these extensive plant breeding trials at least have provided hope that certain traits exist in plant varieties that could be combined by traditional plant breeding or perhaps by genetic engineering techniques to provide "elite" varieties with high levels of resistance to aflatoxin contamination.

Cultural practices such as irrigation are effective in reducing aflatoxin contamination of peanut and corn (Payne, Cassel, and Adkins 1986), but this practice is not always available or cost effective to growers. Other conventional disease control practices, such as the use of fungicides, are largely ineffective in controlling *A. flavus* infection of crops when utilized at concentrations that are cost effective and environmentally safe. Insecticidal control of the pink bollworm is yet another management practice used by cotton growers in Arizona; exit holes in bolls caused by the insect might provide portals of entry for *A. flavus* (Cotty 1989b). Unfortunately, it is not economically feasible to achieve 100% control of the pink bollworm in cotton through the high-frequency use of insecticides, and even relatively low levels of infestation by this insect pest are well correlated to high levels of *A. flavus* infection and subsequent aflatoxin contamination.

Thus, conventional control practices are available that reduce aflatoxin levels in the field, but at substantial, additional cost to the grower. However, the partial effectiveness of these control practices has suggested to researchers that "weak links" exist in the chain of events leading to aflatoxin contamination that could perhaps be exploited even more effectively to interrupt the contamination process.

Nonconventional Methods

Novel methods are needed to control aflatoxin contamination since conventional methods are only partially effective. Permissible levels of aflatoxin and other substances detrimental to animal and human health in foods and feeds are continually being lowered by regulatory agencies and governments worldwide. Therefore, conventional methods probably will not achieve the extremely low levels of aflatoxin to meet regulatory guidelines for sale of food and feed.

Elimination of preharvest aflatoxin contamination might be achieved with novel approaches in biotechnology. To develop new methods to eliminate aflatoxin through biotechnology, additional knowledge is required in three broad areas: (1) ecological and environmental factors affecting aflatoxin production, (2) fundamental molecular and biochemical mechanisms that control the biosynthesis of aflatoxin by the fungus, and (3) biochemistry of host-plant resistance to aflatoxin and aflatoxigenic molds. Knowledge in these areas will aid development of novel methods to interrupt the following chain of events in aflatoxin contamination:

1. The overwintering of the toxigenic fungi and production of primary inoculum
2. Infection of developing fruit and seed by toxigenic fungi
3. Biosynthesis and secretion of aflatoxin into developing and mature seed

This chapter discusses strategies for eliminating preharvest aflatoxin contamination based on our current understanding of the ecological and biochemical processes involved in aflatoxin contamination. Possible control strategies, such as the development of novel biocontrol agents and the development of elite crop lines "immune" to aflatoxin-producing fungi, are discussed.

ECOLOGY OF AFLATOXIGENIC FUNGI

Some basic facts about the ecology and epidemiology of aflatoxigenic fungi have been reviewed (Diener et al. 1987). *A. flavus* and *A. parasiticus* occupy a broad niche in crop environments including the soil, crop debris from the previous growing season, and living tissues of current crops (including cotton bolls/seed and peanut and corn kernels). Of the two aflatoxigenic fungi, *A. flavus* is the predominant cause of aflatoxin contamination of corn and cottonseed, whereas, both *A. parasiticus* and *A. flavus* invade and contaminate peanuts. The broad niche and host range adaptation of these toxigenic fungi must be considered in any approach to destroy or replace them in the field. Since we may never be certain that the entire niche for aflatoxin-producing molds has been identified, generic methods are needed to specifically target the fungi where and when they are active.

CONTROL OF AFLATOXIN CONTAMINATION THROUGH BIOTECHNOLOGY

Three generic approaches for all aflatoxin-susceptible crops could be used to exclude toxigenic fungi from their environmental niches and to regulate aflatoxin biosynthesis:

1. Replace aflatoxigenic strains with nonaflatoxigenic strains in the field (a biocompetitive approach)
2. Incorporate into plant varieties, perhaps through genetic engineering, specific antifungal genes expressed in the specific plant tissues, for example, tissues of the carpel wall of cotton bolls or seed tissues, contaminated by aflatoxigenic strains (a host-plant resistance approach)
3. Inhibit the biosynthetic and secretory process responsible for aflatoxin contamination

This last approach could be the ultimate generic control method since the aflatoxin biosynthetic pathway is the one process that *A. flavus, A. parasiticus,* and all three of the major crops (cotton, corn, and peanut) infected with these fungi share in common.

The Use of Biocompetitive Agents

The use of microbes to control aflatoxin contamination through either their ability to produce antibiotics or their ability to competitively exclude toxigenic fungi has been often suggested (Kimura and Hirano 1988; Wicklow et al. 1987). Technical literature (Cotty 1990) and a recent review of the literature (Cleveland et al. 1990) have pointed out that biocompetitive microbes have advantages over chemical pesticides since they may be more environmentally acceptable, have a longer period of efficacy, and be more readily distributed. Also, biocontrol agents might be effective even when host defenses are weakened during conditions causing high plant stress, which are conditions optimum for aflatoxin contamination (Cleveland et al. 1990). The best biocompetitive agent to control *A. flavus* in the field would be another strain of *A. flavus,* perhaps a nontoxigenic strain of the fungus adapted to that particular field environment and one that maintains aggressiveness and the ability to colonize seed (Cotty 1989a; Cotty 1990). In theory, a biocompetitive agent of the same species as the toxigenic fungus, when properly applied in the field could occupy the same broad ecological niche as toxigenic strains (soil, crop debris, and living host tissues) and ultimately replace toxigenic fungi in field environments. Potential biocompetitive agents consisting of nonaflatoxigenic strains of *A. flavus* were isolated recently from agricultural fields (Cotty, 1989a; Cotty 1990; Joffe 1969). These nontoxigenic strains of *A. flavus* produce little or no aflatoxin while maintaining a high ability to invade plant tissues, thus implying that the aflatoxigenic trait is independent of the ability of toxigenic strains to infect the host plant in the field. Under appropriate conditions, nontoxigenic strains might be able to competitively exclude toxigenic strains, thereby preventing aflatoxin contamination (Cotty 1990). This theory was tested by inoculating developing cotton bolls (Cotty 1990) and corn kernels (Brown, Cotty, and Cleveland 1991) with various aflatoxigenic and non-aflatoxigenic fungal strains through wounds made to simulate insect injury of the crops. When nontoxigenic strains were injected into wounds before inoculation with toxigenic strains, contamination was either prevented or reduced 20- to 100-fold in mature kernels or seed (Cotty 1990; Brown, Cotty, and Cleveland 1991). Simultaneous inoculations also resulted in significant reductions in toxin content of the seed. The results indicated that nontoxigenic strains of *A. flavus* could be biocompetitive agents in the control of aflatoxin contamination. The use of biocompetitive agents is therefore a pragmatic approach in that the best adapted biocompetitive agents (aggressive colonizing strains of *A. flavus*) can be selected for field application to specifically target toxigenic *A. flavus* strains where and when they are active in the field, without necessarily achieving a total understanding of the complex and broad niche occupied by this fungal pest.

Enhancement of Resistance in Plants against Aflatoxigenic Fungi

Plant Resistance Chemicals and Associated Genes

Using conventional plant breeding techniques, researchers have identified varieties of corn demonstrating "resistance" (or, to be more accurate, varieties with less susceptibility) to infection and/or aflatoxin contamination by aflatoxigenic fungi (Lisker and Lillehoj 1991). Identification of chemical compounds linked to this "resistance" would provide biochemical "markers" to monitor during breeding for resistance or in incorporation of the resistance genes into plants by new plant transformation techniques. Advantages and disadvantages must be considered in the conventional plant breeding and new plant transformation technologies for enhancement of host-plant resistance; plant breeding has the advantage of being a known technology, whereas plant transformation techniques have not become completely routine for incorporation of desirable genetic traits/genes into commercially important crops. In addition, plant breeding is an empirical approach that does not depend upon the identification of the biochemical mechanism or function of the trait being sought, and depends only on the ability to screen efficiently for the desirable trait (for example, disease resistance against a particular fungal pest). In contrast, plant transformation technology depends on identification and cloning of the desirable gene for incorporation into the plant genome. However, one of the disadvantages of the plant breeding approach, besides being time-consuming, is that it is often impossible to transfer only the desirable genes of interest into the plant; many times genes of interest are closely linked to a multiple of other genetic traits/genes, perhaps some desirable, some neutral, and some undesirable. The strength of the plant transformation approach is that it can be accomplished relatively quickly (if the technology is available) with only the selected genes of interest, provided they have been identified and cloned. Also, unlike the plant breeding approach, plant transformation technology allows genes to be transferred across the species barrier.

Before resistance traits can be enhanced either by plant breeding or through molecular engineering, specific chemicals linked to resistance must be identified. A further requirement for the genetic engineering approach is that a specific gene(s) for the trait must be identified, cloned, and stably inserted into the plant genome. Potential resistance enzymes and other proteins, and small molecular weight (MW) inhibitory compounds have been identified in crops susceptible to aflatoxin contamination and might be enhanced through plant breeding or genetic engineering. Corn kernels from varieties with varying levels of resistance/susceptibility to *A. flavus*-contained chitinases and glucanases (Neucere, Cleveland, and Dischinger 1991), hydrolytic enzymes often implicated in the lysis of fungal cell walls, and in other plant resistance to fungal pests (Boller 1985). Although potential antifungal enzymes have been identified in corn kernels, a correlation between kernel resistance/susceptibility to *A. flavus* and/or aflatoxin contamination (Lisker and Lillehoj 1991) and levels of hydrolytic antifungal enzymes in the kernels is lacking. Both large MW (McCormick et al. 1988) and small MW compounds (Zeringue and McCormick 1990) were detected in developing cottonseed and in cotton leaves, respectively, that inhibit aflatoxin biosynthesis in *A. flavus* liquid fermentations. In other research, investigators have shown that cotton and peanut contain constitutive levels of antifungal compounds and that under certain conditions these tissues can respond to invading aflatoxigenic molds by producing phytoalexins

(Bell and Stipanovic 1978; Bell 1981; 1983; Cole et al. 1989; Wotton and Strange 1987; Zeringue 1987; Zeringue and McCormick 1989). The discovery of constitutive and *de novo*-produced (phytoalexins) antifungal compounds in crops subject to aflatoxin contamination suggests that endogenous resistance mechanisms exist that could be enhanced through conventional plant breeding or new genetic engineering methods. Exposure of cotton-leaf tissue to the fungus as well as to certain volatile compounds derived from cotton were shown to elicit sesquiterpenoids, such as the cadalenes and their oxidized products, lacinilenes (Zeringue 1987; 1991), which are considered phytoalexins, in tissues remote from the source of these volatiles. Certain volatiles have multiple effects on the physiology of *A. flavus,* ranging from aflatoxin inhibition to aflatoxin stimulation and fungal growth inhibition (Zeringue and McCormick 1989; Zeringue and McCormick 1990). In peanut, small MW phytoalexins, the stilbenes (Cole et al. 1989; Wotton and Strange 1987), are induced by the presence of invading aflatoxigenic fungi, and appear to be correlated with the ability to resist fungal attack.

Genes for some of the potential antifungal hydrolases (chitinases and glucanases) and for the biosynthetic enzymes catalyzing synthesis of certain phytoalexins (e.g., stilbenes) have been cloned (Table 9-1) and could serve as tools in genetic engineering for resistance against toxigenic fungi. For example, the gene for resveratrol synthase, a key enzyme catalyzing biosynthesis of resveratrol (a stilbene phytoalexin), has been cloned from peanut (Lanz, Schroder, and Schroder 1990). Cloning of the resveratrol synthase gene from peanut has important implications in genetic engineering of plants other than peanut, since if this gene is incorporated into other plants, stilbenes could be synthesized from precursors (4-coumaroyl-CoA and malonyl-CoA) commonly available in various plant species. Genes for other key enzymes (e.g., phenylalanine lyase) (Habereder, Schroder, and Ebel 1989) involved in the formation of certain phenylpropanoid precursors of

TABLE 9-1   Potential Fungal Growth Inhibitors from Cotton, Corn (or Other Related Grains), and Peanut

Hydrolases	Lytic Peptides	Phytoalexins
Glucanases[a] (CO,C)	Zeamatins (C)	Stilbenes[b] (P)
Chitinases[a] (P,C)	Thionins[a] (OG)	Lacinilenes (CO)

Note: CO = cotton; C = corn; P = peanut; OG = other grains.
[a]Direct gene products, and some have been cloned from plants.
[b]Not direct gene products; however, the gene for the enzyme (resveratrol synthase) catalyzing synthesis of stilbene phytoalexin in peanut has been cloned.

stilbenes also could be useful in genetic engineering to enhance resistance in peanut, perhaps with genetic engineering techniques to enhance gene expression, such as insertion of multiple gene copies and use of more powerful plant gene promoters or enhancers.

Antifungal hydrolases (chitinases and glucanases) (Kombrink, Schroder, and Hahlbrock 1988) and their genes have been detected in various plant species including crops subject to aflatoxin contamination (Table 9–1): chitinase cDNAs have been cloned from peanut (Herget, Schell, and Schreier 1990), chitinases (Nasser et al. 1988; Neucere, Cleveland, and Dischinger 1991) have been isolated from corn, and glucanases have been detected in cotton (Bucheli et al. 1985) and corn (Jondle, Coors, and Duke 1989).

Other potential fungal growth inhibitors have been detected in crops susceptible to preharvest infection by *A. flavus*. For example, polypeptides (zeamatins or thionins) inhibiting growth of certain fungi have been detected in corn (Vigers, Roberts, and Selitrennikoff 1991) and in related grains (Bohlmann et al. 1988). However, even though hydrolases and polypeptides potentially inhibitory to certain fungi have been detected in aflatoxin-susceptible crops, it is obvious that these substances do not confer complete resistance to aflatoxin-producing fungi, since no truly resistant varieties have been identified. Therefore, it could be speculated that native hydrolases or inhibitory peptides present in aflatoxin-susceptible crops either are not expressed in quantities effective enough to inhibit fungal growth or possess the wrong specificity to efficiently lyse cell walls of aflatoxin-producing fungi. For example, chitinases (Roberts and Selitrennikoff 1988) and inhibitory peptides (Rood, Duvick, and Rao 1991) from different sources have different specificities and effectiveness in inhibiting fungal growth. Future strategies could therefore include cloning and amplifying native antifungal genes or the incorporation of new, nonhost inhibitor genes coding for potent antifungal (against *A. flavus* or *A. parasiticus*) substances into corn, peanut, and cottonseed by genetic engineering techniques.

Molecular Engineering of Antifungal Genes for Tissue-Specific Expression in Plants

Past reports from this and other laboratories have described in detail the biology of the *A. flavus*–host-plant interaction, and this knowledge could be vital to the success of the genetic engineering strategy to incorporate host-plant resistance genes to prevent aflatoxin contamination. For example, *A. flavus* mainly enters cotton bolls through insect injury ports in the carpel wall, such as pink bollworm exit holes (Cotty 1989b), and through natural openings, such as nectaries (Klich and Chmielewski 1985). In a genetic engineering approach, boll entry points involving natural openings might best be guarded against *A. flavus* invasion by inhibitor genes regulated by fungal-induced regulatory sequences, whereas, both wound- and fungal-induced resistance genes might inhibit entry through insect injury portals in the carpel wall. Pertinent examples of genes/gene products induced by pest attack include wound-inducible proteinase inhibitors (in potato) (Cleveland, Thornburg, and Ryan 1987; Thornburg et al. 1987) and fungal-induced peroxidases (in cotton ovule cultures) (Mellon 1985). Although protease inhibitors or peroxidases may not have a direct role in inhibiting fungal growth, regulatory domains of these pest-induced genes could, at least, be useful in the construction of chimeric resistance genes, perhaps using the coding regions of the known antifungal genes previously discussed. Thus, in theory, genetically engineered resistance genes once inserted into the plant genome would be expressed at proper levels, times, and sites in plant tissues to prevent toxigenic fungi from reaching the seed where aflatoxin is produced.

Once outer protective barriers of the boll have been breached, *A. flavus* can grow through the lint to the cottonseed, invade the seed, and produce high levels of aflatoxin. Therefore, wound- or fungal-induced genes producing antifungal growth inhibitors in the carpel wall or locular environment might limit or slow fungal infection of the seed. But as a last resort, expression of antifungal genes in the seed might be necessary to prevent seed invasion by aflatoxin-producing fungi. Several seed-expressed genes, along with their intact regulatory regions, have been cloned from soybean and other plants and shown to be regulated correctly in the seed of transformed tobacco (Goldberg, Barker, and Perez-Grau 1989; Jofuku and Goldberg 1989), thus indicating that nonhost gene regulatory regions could be used to specifically target a specific nonhost tissue site for gene expression. Invasion of cotton bolls by *A. flavus* does not end with the invasion of a single locule. After *A. flavus* invades a specific boll locule, it deteriorates the locular contents and invades the cottonseed, then spreads through intercarpellary septa to adjacent locules. Evidence suggests that intercarpellary septa act as barriers to impede the spread of aflatoxin-producing fungi, at least in strains with relatively low aggressiveness (Cotty 1989*a*). Highly aggressive strains could readily penetrate intercarpellary septa and invade adjacent locules; the only measurable biochemical difference between these high and low aggressive strains is the lack of a major endopolygalacturonase (P2c)—a plant-cell-wall-degrading enzyme—in strains showing low aggressiveness (Cleveland and Cotty 1991). The expression of this polygalacturonase, P2c, may enable strains to spread rapidly between locules, thereby contaminating seed throughout the boll, which suggests the intriguing possibility of genetically engineering cotton to produce polygalacturonase inhibitors (De Lorenzo et al. 1990). In theory, resistance genes (coding for fungal enzyme inhibitors) expressed in the boll tissues or in the intercarpellary septa upon fungal attack might limit the spread of *A. flavus* between locules, thus lowering levels of aflatoxin contamination.

The use of developmentally expressed antifungal genes in the seed might be an appropriate strategy to inhibit aflatoxin contamination of corn, since the defensive barriers in corn kernels (the primary site of attack in the corn plant) might not be responsive to wound- and fungal-induced responses requiring physiologically active tissues, such as the carpel walls of cotton bolls. Similarly, in peanut, either environmentally or developmentally expressed antifungal genes in the peg, pod, testae, and kernel tissues might be effective in limiting invasion by toxigenic fungi into kernels. Little is known about any developmentally expressed antifungal enzymes such as chitinases (Herget, Schell, and Schreier 1990) that exist in peanut kernels or other peanut tissues. However, as pointed out earlier, even if these potentially antifungal enzymes and gene products are not present in levels effective enough to inhibit fungal invasion in peanut, both native and nonnative chitinase genes could be incorporated into peanut or enhanced through new transformation and molecular engineering techniques (Clemente, Weissinger, and Beute 1991).

The phytoalexin response to fungal attack in peanut is a recently documented phenomenon in peanuts (Cole et al. 1989; Wotton and Strange 1987). However, phytoalexins are only expressed in inhibitory levels to fungi when kernel water potential is optimum, and water availability often becomes a limiting factor for maximum expression of kernel resistance during the drought conditions conducive to aflatoxin contamination. Since a key gene involved in stilbene phytoalexin synthesis has been cloned (Lanz, Schroder, and Schroder 1990), perhaps chimeric resistance genes consisting of these genes and selected regulatory DNA sequences that are inducible under lower water potentials could be constructed and stably transformed into the peanut genome, thus endowing kernels with the ability to express the phytoalexin response at low water potentials and during drought conditions.

Transformation of Crops Subject to Aflatoxin Contamination with Antifungal Genes

Perhaps the most difficult technological hurdle that must be overcome in the genetic engineering of crops for resistance to aflatoxin contamination is the development of highly efficient procedures to transform the particular varieties of crops that are subject to aflatoxin contamination with stably expressed foreign genes. Cotton, corn, and peanut, for example, are not as easily transformed as tobacco, which has been the host plant for expression of a variety of foreign genes (Jofuku and Goldberg 1989). Of the three crops subject to aflatoxin contamination, cotton is currently the best model system for testing some of the theories and principles mentioned in this article. Cotton has been successfully transformed, and fully developed cotton plants have been obtained with foreign genes stably expressed in the boll tissues (Firoozabady et al. 1987; Trolinder and Goodin 1988; Umbeck, Swain, and Yang 1989). Therefore, cotton could immediately serve as a host for testing fungal-inhibitor gene constructions. Corn has been successfully but rarely transformed with stably expressed foreign genes (Gordon-Kamm et al. 1990) and could therefore serve only as a future recipient of resistance genes to inhibit infection by *A. flavus*. Peanut has also been transformed with foreign genes (Clemente, Weissinger, and Beute 1991), but successful regeneration of transformed tissue cultures into a fully developed plant with stably-expressed, foreign genes in the fruit has not yet been demonstrated. It should be noted that, unlike dicotyledonous crops, such as cotton, monocotyledonous crops, such as corn, and even certain dicots in the legume family, such as peanut, have been relatively resistant to the transformation and regeneration procedures required to obtain stably expressed foreign genes in fully developed plants.

Interruption of Aflatoxin Biosynthesis

Regulation of Aflatoxin Biosynthesis

As pointed out in previous sections, aflatoxin biosynthesis is not necessarily predestined to occur during fungal growth and infection of seed; numerous examples are available of nonaflatoxigenic strains of *A. flavus* that grow in field and crop environments (Cotty 1989*a*) and even of known aflatoxigenic strains that discontinue production of aflatoxin on certain media or under certain environmental conditions (Lee 1989; Wei and Jong 1986). Also, fungal strains with the potential to produce high levels of aflatoxins have greatly diminished ability to produce toxins in the presence of certain plant metabolites (Bhatnagar and McCormick 1988; McCormick et al. 1988; Zeringue and McCormick 1990). This information indicates that the aflatoxin biosynthetic "machinery" is sensitive to certain genetic, biochemical, and/or environmental influences, and suggests that field strategies could be devised to perturb or interrupt the aflatoxin biosynthetic process. Extensive research efforts are underway in this and other laboratories to understand the molecular basis of aflatoxin biosynthesis so that such strategies could be developed to inhibit aflatoxin contamination of seed. This strategy requires as complete an understanding as possible of the aflatoxin pathway intermediates, enzyme proteins and genes that govern the complex process of aflatoxin biosynthesis.

Aflatoxin biosynthesis involves several biosynthetic steps (Fig. 9–1) requiring several enzymes (Bhatnagar and Cleveland 1990; Bhatnagar, Cleveland, and Kingston 1991; Bhatnagar, Ullah, and Cleveland 1989; Cleveland et al. 1987; Dutton 1988; Singh and Hsieh 1976; Wan and Hsieh 1980; Chuturgoon, Dutton, and Berry, 1990) and genes governing synthesis of these enzymes. The regulation at the gene level for aflatoxin biosynthesis is yet unknown. Most earlier research in the area of the regulation of aflatoxin biosynthesis was accomplished in a broad physiological context without the knowledge of the now isolated pathway intermediates and enzymes. It was discovered, for example, that certain citric acid cycle intermediates influenced aflatoxin biosynthesis (Buchanan et al. 1987). A beginning has now been made in the study of the molecular regulation of aflatoxin synthesis with the assembly of a biosynthetic pathway (Fig. 9–1) (Bennett and Christensen 1983; Bhatnagar, Cleveland, and Lillehoj 1989; Bhatnagar and Cleveland 1990; Bhatnagar, Cleveland, and Kingston 1991; Bhatnagar, Ehrlich and Cleveland 1991) and the identification and partial purification (Singh and Hsieh 1977; Hsieh, Wan, and Billington 1989; Townsend et al. 1988; Yabe, Ando, and Hamasaki 1988; Yabe, Ando, and Hamasaki, 1991; Yabe et al. 1989; 1991; Cleveland and Bhatnagar 1990; 1991; Chuturgoon, Dutton, and Berry 1990) or purification to homogeneity (Bhatnagar, Ullah and Cleveland 1989; Bhatangar, Cleveland and Hsieh 1990; Bhatnagar, Cleveland and Kingston 1991; Keller et al. 1991) of several enzymes catalyzing steps in the pathway.

The regulation of the complex aflatoxin biosynthetic pathway could occur at the transcriptional, translational, or posttranslational levels. In addition, each of the biosynthetic genes could be subject to regulation by regulatory gene products that could act to induce or repress gene transcription. Some enzyme activities catalyzing specific biosynthetic steps might be subject to allosteric regulation by intermediates in the pathway or by other secondary fungal metabolites with a regulatory function. Genes governing aflatoxin biosynthesis could also be regulated by pathway intermediates in a feedback regulation mechanism. Thus, to investigate hypotheses in the molecular regulation of aflatoxin biosynthesis, pertinent pathway genes must be cloned and pathway enzymes must be purified to homogeneity in order to use them as genetic "probes" or biochemical "markers" in model experimental systems.

The major biosynthetic pathway steps between norsolorinic acid (NOR) and aflatoxin B1 in *A. parasiticus* have been reviewed (Bhatnagar, Ehrlich, and Cleveland 1991; Townsend 1986) (Fig. 9–1). Acetate is the basic unit in aflatoxin from experiments that demonstrated incorporation of ^{14}C-labeled acetate into aflatoxin. A multiple-step reaction sequence occurring prior to NOR is as yet hypothetical and is thought to involve a chain of reactions that assembles acetate units into a decaketide backbone (Bhatnagar, Ehrlich, and Cleveland 1991). The polyketide then undergoes a condensation reaction to yield the anthraquinone pigment, NOR. Aflatoxin biosynthesis has been demonstrated to then follow the scheme

NOR→ averantin (AVN)→ averufanin (AVNN)→ averufin (AVR)→ hydroxyversicolorone (HVN)→ versiconal hemiacetal acetate (VHA)→ versicolorin A (VER A)→ demethylsterigmatocystin (DMST)→ sterigmatocystin (ST)→ *O*-methylsterigmatocystin (OMST)→ aflatoxin B1 (AFB1)

(reviewed in Bhatnagar, Ehrlich, and Cleveland 1991).

As recently reviewed (Cleveland and Bhatnagar 1991), elucidation of the biosynthetic steps between NOR and aflatoxin B1 has led to the detection and isolation of some of the aflatoxin pathway enzymes using purified pathway intermediates as substrates. Four aflatoxin pathway enzymes have been purified and have received significant attention in the literature; these enzymes are a reductase (RE), an esterase (ES), a methyltransferase (MT), and an oxidoreductase (OR) catalyzing NOR→ AVN, VHA→ VER A, ST→ OMST, and OMST→ AFB1, respectively (Bhatnagar and Cleveland 1990; Bhatnagar, Cleveland, and Kingston 1991; Bhatnagar, Ullah, and Cleveland 1989; Cleveland et al. 1987; Dutton 1988; Singh and Hsieh 1976; Yabe et al. 1989; Wan and Hsieh 1980; Hsieh, Wan, and Billington 1989; Chuturgoon, Dutton, and Berry 1990). Recently, a desaturase activity converting versicolorin B to versicolorin A has also been identified in fungal extracts (Yabe, Ando, and Hamasaki 1991).

Regulatory studies on the aflatoxin pathway have been initiated (Cleveland and Bhatnagar 1990, 1991; Cleveland et al. 1987, 1991; Hsieh, Wan, and Billington 1989; Chuturgoon and Dutton 1991) using these recently isolated biosynthetic enzymes as biochemical "markers" to monitor the progress of aflatoxin formation in developing *A. parasiticus* cultures. In our laboratory, aflatoxin B1 levels were seen to increase sharply in liquid shake cultures of *A. parasiticus* about 2 days after inoculation (Cleveland and Bhatnagar 1990), at which time mycelial growth rate declines. When RE (unpublished observation in this laboratory; Chuturgoon and Dutton 1991), MT, and OR (Cleveland and Bhatnagar 1990; Cleveland et al. 1987) activities were monitored in the previously mentioned time course studies, they were not detected or were detected in extremely low levels in young (24-hr-old) mycelia, that is, slightly before initiation of aflatoxin synthesis, the activities peaking after nearly 96 hr of incubation in a submerged shake culture. Thus, RE, MT, and OR fit the predicted pattern of specific secondary metabolic enzymes since they are not detected in young, actively metabolizing fungal mycelia, and the activities peak well after fungal growth rate has plateaued. Further investigations on the MT protein in our laboratory have indicated that this enzyme is synthesized *de novo* during the late growth phase of the fungus (Cleveland and Bhatnagar 1990). Therefore, the induction of MT probably is not regulated by a posttranslational activation of a preexisting inactive form of MT. Hence, the appearance of aflatoxin B1 is probably not regulated by the availability of enzyme substrates, but by the appearance of specific pathway enzyme activities. Pathway enzymes are being purified and characterized (Bhatnagar and Cleveland 1991) for the synthesis of antisera and oligonucleotide probes based on purified enzyme proteins so that the genes coding for these enzymes can be isolated and cloned (Bhatnagar and Cleveland 1991; Carey et al. 1992; Keller et al. 1992). Aflatoxin pathway genes can be utilized as probes in studies to determine the molecular regulation of aflatoxin biosynthesis. For example, the gene probes could be used in the detection of specific mRNA transcripts to determine if pathway genes are transcriptionally regulated.

Since RE, ES, MT, and OR activities are not detected (or are detected in low levels) in young mycelia during primary growth, it is likely that these and possibly other enzymes in the aflatoxin pathway are not required for primary metabolic processes, that is, those required for growth of *A. flavus* or *A. parasiticus*. Other evidence that the four enzymes (RE, ES, MT, and OR) are not critical for primary growth is that mutants blocked at the NOR-→AVN, AVN→AVNN, VER A→ST, and OMST→ aflatoxin B1 steps apparently grow in a normal manner in pure culture and on their natural crop substrates (Bhatnagar et al. 1987; Cleveland and Bhatnagar 1990; Cleveland, Bhatnagar, and Brown 1991).

Besides the laboratory mutants that were generated in order to elucidate the biochemistry of aflatoxin synthesis, there are several "natural" aflatoxin non-producing strains that would be valuable for studying the regulation of aflatoxin biosynthesis (Table 9-2). The various categories of aflatoxin-producing and -nonproducing strains were reviewed in a previous article (Cleveland and Bhatnagar 1991). The various strains are classified as follows: (1) those that produce aflatoxin (type I) (Cotty 1990); (2) mutant strains impaired in a particular pathway step (type II) (Bhatnagar et al. 1987; Papa 1982); (3) strains that produce no aflatoxin and accumulate no aflatoxin pathway intermediates, but that apparently possess pathway enzymes; these strains are capable of converting the pathway intermediate, ST, to aflatoxin (type III) (Bennett 1982; Lee 1989); and (4) strains incapable of producing aflatoxin even when supplied with aflatoxin pathway intermediates (type IV) (Bhatnagar, Neucere, and Cleveland 1989). Another category (type V) included in the present review (Table 9-2), in addition to the four reviewed previously (Cleveland and Bhatnagar 1991), consists of fungal strains that possess the ability to carry out most of the steps in the aflatoxin pathway except for the last two (see Fig. 9-1). Type V consists of, for example, strains of *A. nidulans* that produce ST (see Fig. 9-1) as the

TABLE 9-2   Some Categories of *Aspergillus* Species/Strains Based on Presence
or Absence of Aflatoxin or Aflatoxin-Pathway Intermediates and/or Enzymes

Strain Type	Fungal Species	Pathway Enzyme(s) Present or Induced	Pathway Intermediate(s) Present	Aflatoxin Produced
I	F,P	+[a]	+	+
II	F,P	+	+	−/i+[b]
III	F,P	+	−[c]	−/i+
IV	F,P,O	−	−	−
V	N	+/p[d]	+/p	−

Note: F = *A. flavus;* P = *A. parasiticus;* O = *A. oryzae;* N = *A. nidulans.*

[a](+) indicates the pathway where enzymes, intermediates, and/or aflatoxin can be detected.

[b](−/i+) indicates aflatoxin produced if strain is supplied with certain pathway intermediates to bypass enzymes steps inactivated by mutation.

[c](−) indicates the pathway where enzymes, intermediates, and/or aflatoxin could not be detected.

[d](+/p) indicates aflatoxin pathway (enzymes and intermediates) that is "partially" present, but the pathway terminates with the intermediate sterigmatocystin (ST).

final pathway product (Hajjar et al. 1989). The aflatoxin nonproducing strains listed in Table 9-2 could be useful tools for the isolation of genes governing aflatoxin biosynthesis. Certain of the aflatoxin-pathway-blocked mutants included in type II (Table 9-2), for example, have been transformed with native *A. parasiticus* DNA to complement the pathway block (Skory, Chang, and Linz 1991); DNA complementing an aflatoxin pathway block was rescued from a fungal transformant and a putative aflatoxin pathway gene was isolated. Type IV strains are especially interesting, since they may be impaired in a regulatory gene function governing synthesis of several pathway enzymes; the lack of some if not all pathway enzymes in these strains suggests a common regulatory control of biosynthetic enzymes. In theory, utilizing available technology for transformation of *A. flavus* (Woloshuk et al. 1989), regulatory genes could be cloned through transformation, complementation for aflatoxin production, and gene rescue from category IV strains as described previously. Pathway and regulatory genes that govern aflatoxin production will provide the first molecular probes to study the effects of chemical modulators (derived from the fungus or host plant) on the regulation of aflatoxin biosynthesis at the gene level. Future investigations of the biochemistry and molecular genetics of aflatoxin biosynthesis will yield important fundamental information about this process so that strategies can be devised to specifically target and interrupt toxigenesis.

Strategies for Controlling Aflatoxin Based on the "Secondary" Rather than "Primary" Metabolic Roles of Aflatoxin Pathway Enzymes

The frequent reports of aflatoxin nonproducer strains or mutants of *A. flavus* and *A. parasiticus* (types II–IV) (Table 9-2) that grow quite well on a variety of living (plant) and nonliving substrates suggests that aflatoxin pathway enzymes may be "disposable" to the fungus. Therefore, some novel possibilities for the control of aflatoxin in crops are suggested. Some examples include the genetic engineering of *A. flavus* with cloned and inactivated aflatoxin pathway genes in gene replacement techniques to produce stable aflatoxin nonproducing strains of *A. flavus* for use as biocontrol agents. Another possibility would be to utilize analogues of aflatoxin pathway intermediates to inhibit pathway enzymes. Perhaps aflatoxin pathway enzyme inhibitors could be synthesized by a molecular design that affects only specific enzymes involved in aflatoxin synthesis, and not enzyme systems required for primary metabolism of other organisms exposed to the inhibitor chemicals. Using an approach to target the unique fungal enzymes catalyzing aflatoxin biosynthesis, it may be possible to design a new class of ecologically safe pesticides that are specifically inhibitory to aflatoxin synthesis and are nontoxic to plants or to animals consuming pesticide-treated plants.

Additional sources of aflatoxin synthesis inhibitors include several plant-derived, natural product inhibitors (Bhatnagar and McCormick 1988; McCormick et al. 1988; Zaika and Buchanan 1987; Zeringue and McCormick 1989; 1990), with yet unknown mechanisms of action, that could be subject to the same molecular design strategy for development of ecologically safe pesticides. These natural-product inhibitors could also serve as markers for enhancement of aflatoxin resistance traits in plants through classical plant breeding or new genetic engineering techniques. For example, a large MW compound ($\geq$ 10 kdalton) that was isolated from cottonseed coats demonstrated very little effect on fungal growth, but effectively reduced aflatoxin production in fungal culture (McCormick et al. 1988). Other examples of natural plant products from a variety of plant sources that inhibit aflatoxin production include compounds in extracts from the neem plant (Bhatnagar and McCormick 1988; Zeringue and Bhatnagar 1990) and several other plant products that have varying effects on growth of aflatoxigenic fungi and on aflatoxin production (Zaika and Buchanan 1987; Zeringue and McCormick 1989; Zeringue and McCormick 1990).

CONCLUSION

Aflatoxin contamination of important food commodities is a unique problem in agriculture unlikely to be completely solved by conventional control strategies routinely used against other fungal pests of plants. Current control practices to eliminate fastidious fungal parasites of plants such as those involved in race specific interactions are considerably less effective in controlling facultative pathogens such as *A. flavus* and *A. parasiticus*. Perhaps the difficulty in developing effective control measures for aflatoxin-producing fungi stems from the difficulty in eliminating these fungi from an extremely broad niche. Therefore, a more targeted approach to eliminate key steps in the infection and/or aflatoxin biosynthetic process might be more effective in reducing or eliminating aflatoxin contamination. For example, incorporation of the genetic potential into plants for hydrolyzing specific components of fungal cell walls or in interrupting the aflatoxin biosynthetic pathway (a host plant modification approach) and/or replacing toxigenic strains in the field with "domesticated" nontoxigenic strains of the same fungal species (a fungal modification approach) are viable approaches for elimination of aflatoxin contamination through biotechnology.

References

Bell, A. A. 1981. Biochemical mechanisms of disease resistance. *Ann. Rev. Plant Physiol.* **32**:21–81.

Bell, A. A. 1983. Morphology, chemistry, and genetics of *Gossypium* adaptations to pests. In *Phytochemical Adaptions to Stress, Recent Advances in Phytochemistry,* Vol. 18, ed. B. N. Timmermann, C. Steelink, and F. A. Loewus. pp. 197–229. New York: Plenum.

Bell, A. A., and R. D. Stipanovic. 1978. Biochemistry of disease and pest resistance in cotton. *Mycopathologia* **65**:91–106.

Bennett, J. W. 1982. Genetics of mycotoxin production with emphasis on aflatoxins. In *Overproduction of Microbial Products,* ed. V. Krumphanzl, B. Sikyta, and Z. Vanek, pp. 549–561. New York: Academic Press.

Bennett, J. W., and S. B. Christensen. 1983. New perspectives on aflatoxin biosynthesis. *Adv. Appl. Microbiol.* **19**:53–92.

Bhatnagar, D., and T. E. Cleveland. 1990. Purification and characterization of a reductase from *Aspergillus parasiticus* SRRC 2043 involved in aflatoxin biosynthesis. *FASEB J.* **4**:2727.

Bhatnagar, D., and T. E. Cleveland. 1991. Aflatoxin biosynthesis. Developments in chemistry, biochemistry and genetics. In *Aflatoxin in Corn, New Perspectives,* eds. O. L. Shotwell, and C. R. Hurburg, Jr., pp. 391–405. Ames, Iowa: Iowa State University.

Bhatnagar, D., and S. P. McCormick. 1988. The inhibitory effect of neem (*Azadirachata indica*) leaf extracts on aflatoxin synthesis in *Aspergillus parasiticus*. *J. Am. Oil Chemists' Soc.* **65**:1166–1168.

Bhatnagar, D., T. E. Cleveland, and D. G. I. Kingston. 1991. Enzymological evidence for separate pathways for aflatoxin B_1 and B_2 biosynthesis. *Biochemistry* **30**:4343–4350.

Bhatnagar, D., T. E. Cleveland, and E. B. Lillehoj. 1989. Enzymes in late stages of aflatoxin B_1 biosynthesis: Strategies for identifying pertinent genes. *Mycopathologia* **107**:75–83.

Bhatnagar, D., K. C. Ehrlich, and T. E. Cleveland. 1991. Oxidation-reduction reactions in biosynthesis of secondary metabolites. In *Handbook of Applied Mycol-*

ogy, Vol. 5, *Mycotoxins in Ecological Systems,* ed. D. Bhatnagar, E. B. Lillehoj, and D. K. Arora, pp. 255-286. New York: Marcel Dekker.

Bhatnagar, D., J. N. Neucere, and T. E. Cleveland. 1989. Immunochemical detection of aflatoxigenic potential of *Aspergillus* species with antisera perpared against enzymes specific to aflatoxin biosynthesis. *Food Agric. Immunol.* **1:**225-234.

Bhatnagar, D., A. H. J. Ullah, and T. E. Cleveland. 1989. Purification and characterization of a methyltransferase from *Aspergillus parasiticus* SRRC 163 involved in aflatoxin biosynthetic pathway. *Preparative Biochem.* **18:**321-349.

Bhatnagar, D., S. P. McCormick, L. S. Lee, and R. A. Hill. 1987. Identification of O-methylsterigmatocystin as an aflatoxin B_1 and G_1 precursor in *Aspergillus parasiticus. Appl. Environ. Microbiol.* **53:**1028-1033.

Bohlmann, H., S. Clausen, S. Behnke, H. Giese, C. Hiller, U. Reimann-Philipp, G. Schrader, V. Barkholt, and K. Apel. 1988. Leaf-specific thionins of barley-a novel class of cell wall proteins toxic to plant-pathogenic fungi and possibly involved in the defense mechanism of plants. *EMBO J.* **7:**1559-1565.

Boller, T. 1985. Induction of hydrolases as a defense reaction against pathogens. In *Cellular and Molecular Biology of Plant Stress,* ed. J. L. Key and T. Kosuge, pp. 247-262. New York: A. R. Liss.

Bressac, B., M. Kew, J. Wands, and M. Ozturk. 1991. Selective G to T mutations of p53 gene in hepatocellular carcinoma from southern Africa. *Nature* **350:**429-431.

Brown, R. L., P. J. Cotty, and T. E. Cleveland. 1991. Reduction in aflatoxin content of maize by atoxigenic strains of *Aspergillus flavus. J. Food Prot.* **54:**623-625.

Buchanan, R. L., S. B. Jones, W. V. Gerasimowicz, L. L. Zaika, H. G. Stahl, and L. A. Ocker. 1987. Regulation of aflatoxin biosynthesis: Assessment of the role of cellular energy status as a regulator of the induction of aflatoxin production. *Appl. Environ. Microbiol.* **53:**1224-1231.

Bucheli, P., M. Durr, A. J. Buchala, and H. Meier. 1985. Beta glucanases in developing cotton *Gossypium hursutum* fibers. *Planta* **166:**530-536.

Campbell, T. C., J. Chen, C. Liu, J. Li, and B. Parpia. 1990. Nonassociation of aflatoxin with primary liver cancer in a cross-sectional ecological survey in the People's Republic of China. *Cancer Res.* **50:**6882-6893.

Cary, J. W., T. E. Cleveland, and D. Bhatnagar. 1992. Regulation by thiamine of expression of a gene from *Aspergillus parasiticus* encoding norsolorinic acid reductase activity. *FASEB J.* **5:**A288.

Chuturgoon, A. A., and M. F. Dutton. 1991. The appearance of an enzyme activity catalyzing the conversion of norsolorinic acid to averantin in *Aspergillus parasiticus* cultures. Mycopathologia **113:**41-44.

Chuturgoon, A. A., M. F. Dutton, and R. K. Berry. 1990. The preparation of an enzyme associated with aflatoxin biosynthesis by affinity chromatography. *Biochem. Biophys. Res. Comm.* **166:**38-42.

Clemente, T. E., A. K. Weissinger, and M. K. Beute. 1991. Stable transgenic peanut (*Arachis hypogeae* L.) calli produced by high velocity microprojectile bombardment. *Phytopathology* **81:**1170.

Cleveland, T. E., and D. Bhatnagar. 1990. Evidence for de novo synthesis of an aflatoxin pathway-methyltransferase near the cessation of active growth and the onset of aflatoxin biosynthesis by *Aspergillus parasiticus* mycelia. *Can. J. Microbiol.* **36**:1–5.

Cleveland, T. E., and D. Bhatnagar. 1991. Molecular regulation of aflatoxin biosynthesis. In *Mycotoxins, Cancer and Health,* Pennington Center Nutrition Series, Vol. 1, ed. G. A. Bray and D. H. Ryan, pp. 270–287. Baton Rouge, LA: Louisiana State University Press.

Cleveland, T. E., and P. J. Cotty. 1991. Invasiveness of *Aspergillus flavus* isolates in wounded cotton bolls is associated with production of a specific fungal polygalacturonase. *Phytopathology* **81**:155–158.

Cleveland, T. E., D. Bhatnagar, and R. L. Brown. 1991. Aflatoxin production via cross-feeding of pathway intermediates during co-fermentation of aflatoxin pathway-blocked mutants of *Aspergillus parasiticus. Appl. Environ. Microbiol.* **57**:2907–2911.

Cleveland, T. E., R. W. Thornburg, and C. A. Ryan. 1987. Molecular characterization of a wound-inducible inhibitor I gene from potato and the processing of its mRNA and protein. *Plant Mol. Biol.* **8**:199–207.

Cleveland T. E., A. R. Lax, L. S. Lee, and D. Bhatnagar. 1987. Appearance of enzyme activities catalyzing conversion of sterigmatocystin to aflatoxin B_1 in late-growth-phase *Aspergillus parasiticus* cultures. *Appl. Environ. Microbiol.* **53**:1711–1713.

Cleveland, T. E., D. Bhatnagar, P. J. Cotty, and H. J. Zeringue. 1990. Aflatoxin reduction—A molecular strategy. In *Biotechnology and Food Safety,* ed. D. D. Bill and S.-D. Kung, pp. 117–135. Stoneham, MA: Butterworth-Heine Meun.

Cleveland, T. E., J. W. Cary, D. Bhatnagar, and N. P. Keller. 1991. Regulation of norsolorinic acid reductase, a key enzyme catalyzing an early step in aflatoxin B1 biosynthesis by *Aspergillus parasiticus. Phytopathology* **81**:1219–1220.

Cole, R. J., T. H. Sanders, J. W. Dorner, and P. D. Blankenship. 1989. Environmental conditions required to induce preharvest aflatoxin contamination of groundnuts: Summary of six years research. In *Aflatoxin Contamination of Groundnuts,* pp. 279–287. Patancheru, India: International Crop Research Institute for Semi-Arid Tropics.

Cotty, P. J. 1989*a*. Virulence and cultural characteristics of two *Aspergillus flavus* strains pathogenic on cotton. *Phytopathology* **79**:808–814.

Cotty, P. J. 1989*b*. Aflatoxin contamination of cottonseed: Comparison of pink bollworm damaged and undamaged bolls. *Tropical Sci.* **29**:273–277.

Cotty, P. J. 1990. Effect of atoxigenic strains of *Aspergillus flavus* on aflatoxin contamination of developing cottonseed. *Plant Dis.* **74**:233–235.

De Lorenzo, G., J. Ito, R. D'Ovidio, F. Cervone, P. Albersheim, and A. G. Darvill. 1990. Host-pathogen interactions. XXXVII. Abilities of the polygalacturonase-inhibiting proteins from four cultivars of *Phaseolus vulgaris* to inhibit the endo-polygalacturonases from three races of *Colletotrichum lindemuthianum. Physiol. Mol. Plant Pathol.* **36**:421–435.

Diener, U. L., R. J. Cole, T. H. Sanders, G. A. Payne, L. S. Lee, and M. A. Klich. 1987. Epidemiology of aflatoxin formation by *Aspergillus flavus. Ann. Rev. Phytopathol.* **25**:249–270.

Dutton, M. F. 1988. Enzymes and aflatoxin biosynthesis. *Microbiol. Rev.* **52**:274–295.

Firoozabady, E., D. L. DeBoer, D. J. Merlo, E. L. Halk, L. N. Amerson, K. E. Rashka, and E. E. Murray. 1987. Transformation of cotton (*Gossypium hirsutum* L.) by *Agrobacterium tumefaciens* and regeneration of transgenic plants. *Plant Mol. Biol.* **10**:105–116.

Goldberg, R., S. J. Barker, and L. Perez-Grau. 1989. Regulation of gene expression during plant embryogenesis. *Cell* **56**:149–160.

Gordon-Kamm, W. J., T. M. Spencer, M. L. Mangano, T. R. Adams, R. J. Daines, W. G. Start, J. V. O'Brien, S. A. Chambers, W. R. Adams, N. G. Willetts, T. B. Rice, C. J. Mackey, R. W. Krueger, A. P. Kausch, and P. G. Lemaux. 1990. Transformation of maize cells and regeneration of fertile transgenic plants. *Plant Cell* **2**:603–618.

Gorman, D. P., and M. S. Kang. 1991. Preharvest aflatoxin contamination in maize: Resistance and genetics. *Plant Breeding* **107**:1–10.

Groopman, J. D., and G. Sabbioni, 1991. Detection of aflatoxin and its metabolites in human biological fluids. In *Mycotoxins, Cancer and Health,* Pennington Center Nutrition Series, Vol. 1, ed. G. A. Bray and D. H. Ryan, pp. 18–31. Baton Rouge, LA: Lousiana State University Press.

Habereder, H., G. Schroder, and J. Ebel. 1989. Rapid induction of phenylalanine ammonia-lyase and chalcone synthase mRNAs during fungus infection of soybean (*Glycine max* L.) roots or elicitor treatment of soybean cell cultures at the onset of phytoalexin synthesis. *Planta* **177**:58–65.

Hajjar, J. D., J. W. Bennett, D. Bhatnagar, and R. Bahu. 1989. Sterigmatocystin production by laboratory strains of *Aspergillus nidulans. Mycol. Res.* **93**:548–551.

Herget, T., J. Schell, and P. H. Schreier. 1990. Elicitor-specific induction of one member of the chitinase gene family in *Arachis hypogaea. Mol. Gen. Genet.* **224**:469–476.

Hsieh, D. P. H., C. C. Wan, and J. A. Billington. 1989. A versiconal hemiacetal acetate converting enzyme in aflatoxin biosynthesis. *Mycopathologia* **107**:121–126.

Hsu, I. C., R. A. Metcalf, T. Sun, J. A. Welsh, N. J. Wang, and C. C. Harris. 1991. Mutational hotspot in the p53 gene in human hepatocellular carcinomas. *Nature* **350**:427–428.

Jelinek C. F., A. E. Pohland, and G. E. Wood. 1989. Review of mycotoxin contamination. Worldwide occurrence of mycotoxins in foods and feeds—an update. *J. Assoc. Off. Anal. Chemists* **72**:223–230.

Joffe, A. Z. 1969. Aflatoxin produced by 1626 isolates of *Aspergillus flavus* from groundnut kernels and soils in Israel. *Nature* **221**:492.

Jofuku, K. D., and R. B. Goldberg. 1989. Kunitz trypsin inhibitor genes are differentially expressed during the soybean life cycle and in transformed tobacco plants. *Plant Cell* **1**:1079–1093.

Jondle, D. J., J. G. Coors, and S. H. Duke. 1989. Maize leaf beta-1,3 glucanase activity in relation to resistance to *Exserohilum turcicum. Can. J. Bot.* **67**:263–266.

Keller, N. P., G. A. Payne, T. E. Cleveland, and D. Bhatnagar. 1991. Unique DNAs distinguish *Aspergillus flavus* and *A. parasiticus* strains. *Phytopathology* **81**:1240.

Keller, N. P., H. C. Dischinger, D. Bhatnagar, T. E. Cleveland, and A. H. J. Ullah. 1992. Purification of a second methyltransferase active in the aflatoxin biosynthetic pathway. *Biochem. Biophys. Acta* (in press).

Kimura, N., and Hirano, S. 1988. Inhibitory strains of *Bacillus subtillis* for growth and aflatoxin-production of aflatoxigenic fungi. *Agric. Biol. Chem.* **52**:1173–1179.

Klich, M. A., and Chmielewski, M. A. 1985. Nectaries as entry sites for *Aspergillus flavus* in developing cotton bolls. *Appl. Environ. Microbiol.* **50**:602–604.

Kombrink, E., M. Schroder, and K. Hahlbrock. 1988. Several "pathogenesis-related" proteins in potato are 1,3-β-glucanases and chitinases. *Proc. Natl. Acad. Sci. (USA)* **85**:782–786.

Lanz, T., G. Schroder, and J. Schroder. 1990. Differential regulation of genes for resveratrol synthase in cell cultures of *Arachis hypogaea* L. *Planta* **181**:169–175.

Lee, L. S. 1989. Metabolic precursor regulation of aflatoxin formation in toxigenic and non-toxigenic strains of *Aspergillus flavus. Mycopathologia* **107**:127–130.

Lisker, N., and E. B. Lillehoj. 1991. Prevention of mycotoxin contamination (principally aflatoxins and *Fusarium* toxins) at the preharvest stage. In *Mycotoxins and Animal Foods,* ed. J. E. Smith and R. S. Henderson, pp. 689–719. Boca Raton, FL: CRC Press.

McCormick, S. P., D. Bhatnagar, W. R. Goynes, and L. S. Lee. 1988. An inhibitor of aflatoxin biosynthesis in developing cottonseed. *Can. J. Bot.* **66**:998–1002.

Mellon, J. E. 1985. Elicitation of cotton isoperoxidases by *Aspergillus flavus* and other fungi pathogenic to cotton. *Physiol. Plant Pathol.* **27**:280–288.

Nasser, W., M. de Tapia, S. Kauffmann, S. Montasser-Konhsari, and G. Burkard. 1988. Identification and characterization of maize pathogenesis-related proteins. Four maize PR proteins are chitinases. *Plant Mol. Biol.* **11**:529–538.

Neucere, J. N., T. E. Cleveland, and C. Dischinger. 1991. Existence of chitinases in mature corn kernels (*Zea mays* L.). *J. Agric. Food Chem.* **39**:1326–1328.

Papa, K. E. 1982. Norsolorinic acid mutant of *Aspergillus flavus. J. Gen. Microbiol.* **128**:1345–1348.

Park, D. L., L. S. Lee, R. L. Price, and A. E. Pohland. 1988. Review of the decontamination of aflatoxins by ammoniation: Current status and regulation. *J. Assoc. Off. Anal. Chemists* **71**:685–703.

Payne, G. A., D. K. Cassel, and C. R. Adkins. 1986. Reduction of aflatoxin contamination in corn due to irrigation and tillage. *Phytopathology* **76**:679–684.

Phillips, T., B. A. Sarr, B. A. Clement, L. F. Kubena, and R. B. Harvey. 1991. Prevention of aflatoxicosis in farm animals via selective chemisorption of aflatoxin. In *Mycotoxins, Cancer and Health,* Pennington Center Nutrition Series, Vol. 1, ed. D. H. Ryan and G. A. Bray, pp. 223–237. Baton Rouge, LA: Louisiana State University Press.

Roberts, W. K., and C. P. Selitrennikoff. 1988. Plant and bacterial chitinases differ in antifungal activity. *J. Gen. Microbiol.* **134**:169–176.

Rood, T. A., J. P. Duvick, and A. G. Rao. 1991. *In vitro* activity of selected proteins and peptides against plant pathogenic fungi. *Phytopathology* **81**:1196.

Scott, G. E., N. Zummo, E. B. Lillehoj, N. W. Widstrom, M. S. Kang, D. R. West, G. A. Payne, T. E. Cleveland, O. H. Calvert, and B. A. Fortnum. 1991. Aflatoxin in corn hybrids field inoculated with *Aspergillus flavus. Agronomy J.* **83**:595–598.

Singh, R., and D. P. H. Hsieh. 1976. Enzymatic conversion of sterigmatocystin into aflatoxin B₁ by cell-free extracts of *Aspergillus parasiticus. Appl. Environ. Microbiol.* **31**:743–745.

Singh, R., and D. P. H. Hsieh. 1977. Aflatoxin biosynthetic pathway: Elucidation by using blocked mutants of *Aspergillus parasiticus. Arch. Biochem. Biophys.* **178**:285–292.

Skory, C. D., P. K. Chang, and J. Linz. 1991. Isolation of a gene involved in the aflatoxin biosynthetic pathway, Abstract O-11. In *Proceedings of the 91st General Meeting of the American Society for Microbiology.* Dallas, Texas.

Thornburg, R. W., G. An, T. E. Cleveland, R. Johnson, and C. A. Ryan. 1987. Wound-inducible expression of potato inhibitor II gene in transgenic tobacco plants. *Proc. Natl. Acad. Sci. (USA)* **84**:744–748.

Townsend, C. A. 1986. Progress towards a biosynthetic rationale of the aflatoxin pathway. *Pure Appl. Chem.* **58**:227–238.

Townsend, C. A., K. A. Plaucan, K. Pal, and S. W. Brobst. 1988. Hydroxyversicolorone: Isolation and characterization of a potential intermediate in aflatoxin biosynthesis. *J. Org. Chem.* **53**:2472–2477.

Trolinder, N. L., and J. R. Goodin. 1988. Somatic embryogenesis in cotton (*Gossypium*). II. Requirements for embryo development and plant regeneration. *Plant Cell, Tissue and Organ Culture* **12**:43–53.

Umbeck, P., W. Swain, and N.-S. Yang. 1989. Inheritance and expression of genes for kanamycin and chloramphenicol resistance in transgenic cotton plants. *Crop Sci.* **29**:196–201.

Vigers, A. J., W. K. Roberts, and C. P. Selitrennikoff. 1991. A new family of plant antifungal proteins. *Mol. Plant-Microbe Interactions* **4**:315–323.

Wan, N. C., and D. P. H. Hsieh. 1980. Enzymatic formation of the bisfuran structure in aflatoxin biosynthesis. *Appl. Environ. Microbiol.* **39**:109–112.

Wei, D.-L., and S.-C. Jong. 1986. Production of aflatoxin by strains of the *Aspergillus flavus* group maintained in ATCC. *Mycopathologia* **93**:19–24.

Wicklow, D. T., B. W. Horn, O. L. Shotwell, C. W. Hesseltine, and R. W. Caldwell. 1987. Fungal interference in *Aspergillus flavus* infection and aflatoxin contamination of maize grown in a controlled environment. *Phytopathology* **78**:68–74.

Widstrom, N. W. 1987. Breeding strategies to control aflatoxin contamination of mazie through host plant resistance. In *Aflatoxin in Maize: A Proceedings of the Workshop,* ed. M. S. Zuber, E. B. Lillehoj, and B. L. Renfro, pp. 212–220. Mexico, D. F.: CIMMYT.

Wogan, G. N. 1991. Aflatoxins as risk factors for primary hepatocellular carcinoma in humans. In *Mycotoxins, Cancer and Health,* Pennington Center Nutrition

Series, Vol. 1, ed. G. A. Bray and D. H. Ryan, pp. 3-17. Baton Rouge, LA: Louisiana State University Press.

Woloshuk, C. P., E. R. Seip, G. A. Payne, and C. R. Adkins. 1989. Genetic transformation system for the aflatoxin-producing fungus *Aspergillus flavus. Appl. Environ. Microbiol.* **55**:86-90.

Wotton, H. R., and R. N. Strange. 1987. Increased susceptibility and reduced phytoalexin accumulation in drought-stressed peanut kernels challenged with *Aspergillus flavus. Appl. Environ. Microbiol.* **53**:270-273.

Yabe, K., Y. Ando, and T. Hamasaki. 1991. Desaturase activity in the branching step between aflatoxin B_1 and G_1 and aflatoxin B_2 and G_2. *Agric. Biol. Chem.* **55**:1907-1911.

Yabe, K., Y. Ando, J. Hashimoto, and T. Hamasaki. 1989. Two distinct O-methyltransferases in aflatoxin biosynthesis. *Appl. Environ. Microbiol.* **55**:2172-2177.

Yabe, K., Y. Nakamura, H. Nakajima, Y. Ando, and T. Hamasaki. 1991. Enzymatic conversion of norsolorinic acid to averufin in aflatoxin biosynthesis. *Appl. Environ. Microbiol.* **57**:1340-1345.

Yeh, F.-S., M. C. Ju, C.-C. Mo, S. Lou, M. J. Tong, and B. E. Henderson. 1989. Hepatitis B virus, aflatoxins, and hepatocellular carcinoma in Southern Guangxi, China. *Cancer Res.* **49**:2506-2709.

Zaika, L. L., and R. L. Buchanan. 1987. Review of compounds affecting the biosynthesis or bioregulation of aflatoxins. *J. Food Prot.* **50**:691-708.

Zeringue, H. J. 1987. Changes in cotton leaf chemistry induced by volatile elicitors. *Phytochemistry* **26**:1357-1360.

Zeringue, H. J. 1991. Effect of C_6-C_{10} alkenals and alkanals on eliciting a defense response in the developing cotton boll. *Phytochemistry* (in press).

Zeringue, H. J., and D. Bhatnagar. 1990. Inhibition of aflatoxin production in *Aspergillus flavus* infected cotton bolls after treatment with neem (*Azadirachta indica*) leaf extracts. *J. Am. Oil Chem. Soc.* **67**:215-216.

Zeringue, H. J., and S. P. McCormick. 1989. Relationships between cotton leaf-derived volatiles and growth of *Aspergillus flavus. J. Am. Oil Chemists' Soc.* **66**:581-585.

Zeringue, H. J., and S. P. McCormick. 1990. Aflatoxin production in cultures of *Aspergillus flavus* incubated in atmospheres containing selected cotton leaf-derived volatiles. *Toxicon* **28**:445-448.

10

Molecular Strategies for Improving the Quality of Muscle Food Products

Arthur M. Spanier and Peter B. Johnsen

The flavor quality of muscle foods is dependent upon several key premortem factors such as an animal's age, breed, sex, and nutritional status. While these factors are all involved in the development of the final flavor of a food, additional textural and flavor changes develop both during the period of postmortem aging (Koohmaraie et al. 1988; Ouali 1990) and during cooking and storage of the food (Etherington, Taylor, and Dansfield 1987; Kato and Nishimura 1987; Ouali et al. 1987; Spanier, Edwards, and Dupuy 1988; St. Angelo et al. 1987, 1988).

Muscle foods show significant alteration in the type and level of chemical components, such as sugars, organic acids, peptides and free amino acids, and products of adenine nucleotide metabolism, for example, adenosine triphosphate (ATP), during the postmortem conditioning period and during cooking/storage. Many of these changes are due to enhanced hydrolytic activity (Spanier, McMillin, and Miller 1990). Regardless of the history of the meat, heating develops both desirable and undesirable flavors from the thermal degradation and the interaction of sugars, amino acids, proteins, nucleotides, Maillard reactions, and lipid oxidation. Thus, the chemical modifications that occur during postmortem aging and during the subsequent handling of meat serve as a pool of reactive flavor compounds and flavor intermediates that later interact to form additional flavor notes during cooking, for example, sugars and amino acids react during heating to form Maillard products (Bailey 1988; Bailey et al. 1987). It is apparent, therefore, that the development of flavor in meat is an extremely complicated process that occurs continuously from before slaughter through cooking and storage, and ending when the food is eaten and the flavor perceived. The final quality of meat therefore involves both external and internal factors. These factors, when combined, establish flavor quality and influence the purchasing decision of the consumer. It is thus essential that the food technologist become cognizant of the molecular mechanisms of food flavor development and deterioration so that molecular strategies can be developed that will positively impact meat flavor quality.

IMPROVING MUSCLE FOOD QUALITY: PREHARVEST

Transgenic Animals

The practice of gene transfer between animal species promises tremendous benefits for agriculture and the food industry (Pursel et al. 1989). Several approaches to respond to consumer needs have been used in the production of agriculturally important muscle foods and their products.

A *transgenic animal* is an animal that has had a gene sequence or sequences inserted into its genome by laboratory techniques. In the animal kingdom, the most common gene insertion technique is to inject the transgene-DNA into the pronucleus of a single-celled embryo where it integrates into the embryo's genome. As many as 25% of the transgenic embryos implanted into host mothers result in the birth of transgenic animals (Westphal 1989). The new animal will have the new trait, which will be propagated by passage from parent to new progeny.

The genetic engineering of livestock has been initiated and accomplished to a limited degree (Table 10-1). For example, Pursel et al. (1987) fused human and bovine growth hormone genes to the metallothionein (MT) promoter and introduced this gene for growth hormone into the germ line of sheep and pigs. According to Westphal (1989), other groups have produced giant fertile pigs with increased body weight and an impressive reduction in back fat by constructing a transgene of bovine growth hormone fused to a regulatory region of rat *P*-enolpyruvate carboxykinase (PEPCK) gene. Expression in the latter case was specific for the liver and kidney, but most importantly, the investigators were able to regulate the transgene by altering the levels of dietary carbohydrate and proteins in a manner that was predicted by studies of the genuine PEPCK gene. Also, fish, which are rapidly becoming a significant agricultural crop through aquaculture, are subjects of genetic manipulation, 13 species being involved to date (Kapuscinski and Hallerman 1990a).

While this technology has a great potential as a source of a new medicines and foods, real and perceived risks need to be addressed. One simpler minor risk is the possibility that the desired trait will not be expressed even though it has been inserted into the genome of the host animal. Furthermore, while a gene may be inserted and expressed by the host animal, there

TABLE 10-1   Production of Transgenic Farm Animals[a]

Gene	Species	Result Integration	Result Expression
ALV	Chicken	✔	
REV	Chicken	✔	
ALV	Chicken		✔
BPV	Cow	✔	
mMT/hGH	Fish		✔
cα-Crystallin	Fish		✔
hGH	Fish	✔	
mMT/hGH	Fish	✔	
mMT/βGal	Fish		✔
SV/hygro	Fish	✔	
mMT/hGH	Pig	✔	✔
mMT/bGH	Pig		✔
hMT/pGH	Pig		✔
MLV/rGH	Pig		✔
bPRL/bGH	Pig		✔
mMT/hGH	Rabbit	✔	✔
hMT/hGH	Rabbit		✔
rbEμ/rbc-myc	Rabbit		✔
mMT/hGH	Sheep	✔	
mMT/TK	Sheep	✔	
oβLG/hFIX	Sheep	✔	
oβLG/hα1AT	Sheep	✔	
mMT/bGH	Sheep		✔
mMT/hGRF	Sheep		✔
oMT/oGH	Sheep		✔
oβLG/hFIX	Sheep		✔

Source: Adapted from Pursel et al. 1989.

[a]Listed are genes that have been introduced into six species of animals. The slash separates the promoter or enhancer of one gene from the structural gene. The species is indicated by a lower case letter before the abbreviation of the gene: b, bovine; c, chicken; h, human; m, mouse; o, ovine; p, porcine; r, rat; rb, rabbit. Gene abbreviations: ALV, avian leukosis virus; α1AT, α1 anti-trypsin; BPV, bovine papilloma virus; Eμ, immunoglobulin heavy chain; FIX, factor IX; GH, growth hormone; βGal, galactosidase; hygro, hygromycin; βLG, β-lactoglobulin; MT, metallothionein; MLV, Moloney murine leukemia virus; REV, reticuloendotheliosis; PRL, prolactin; SV, SV40; TK, thymidine kinase. Integration means the gene became part of the DNA complement; expression indicates that the gene was integrated, and either transgene mRNA or protein (or both) was detectable.

can be difficulty in turning the gene on and off at the appropriate time. Although the gene or trait is engineered and expressed by the host, the potential always exists for loss of this trait in succeeding generations, prompting the suggestion that multigenerational studies must be performed to fully understand the relative value of a transgenically modified animal (Pursel et al. 1989).

More complex is the potential impact the transferred gene or its product will have on the general well-being of the host animal, that is, any secondary effects the trait may have on the normal physiological function of the recipient of the gene. Recent investigations have indicated that such secondary effects are likely. Pursel et al. (1989) have shown that although several successive generations of pigs expressing the bovine growth hormone (bGH) gene showed improvement in daily weight gain and feeding efficiency, and exhibited a marked reduction in subcutaneous fat, they also showed some long-term detrimental effects from the inserted bGH gene, including the expression of a high incidence of "gastric ulcers, arthritis, cardiomegaly, dermatitis, and renal disease."

Another major area of public concern regarding transgenic animals is the potential impact these animals, once released, might have on the environment. For example, could the transformed animal overgraze a field or will the engineered organism out-compete native organisms in the environment for available food sources? The American Fisheries Society has raised this concern for transgenic fishes, and proposed support of restrictions on the use of transgenic fishes in aquaculture, pending completion of the following: (1) a thorough program of risk assessment, and (2) demonstration of minimal environmental risk (Kapuscinski and Hallerman 1990b). The popular press has tended to promote heightened public expectations, and fisheries resources managers responsible for wild and native species are concerned that public expectations for transgenic fishes will make their task more difficult. However, proponents of the use of transgenic species for agriculture express little concern that "corn or cows gone awry will spread as irreversibly as the ice-minus bacteria" developed to protect plants from frost (Shotwell 1991).

Biotechnological enhancement of food animals is not restricted to genetic alteration of the animal itself, but can involve the administration of bioengineered gene products to manipulate metabolic function. The two most widespread examples are BST (bovine somatotropin) and PST (porcine somatotropin). BST, a protein hormone manufactured through recombinant DNA technology, increases milk production per cow 10–15% while utilizing 5–10% less food per unit of milk, thereby reducing milk production costs (Fallert et al. 1987). PST, used to improve feed efficiency, can increase lean muscling and reduce fat deposition. Hogs administered PST demonstrate 10–45% increases in rate of weight gain, 15–35% improved feed efficiency, 15–70% reduction of back fat, and a 10–50% increase in loin size (Hayenga 1988).

Public concern has been expressed concerning the potential effect that the bioengineered gene might have on the consumer if the transgenically altered products are used for human food, particularly when the gene product is a hormone or bioactive peptide. Could such physiologically active substances affect the recipient's metabolism or normal body function? Two recent investigations (Westphal 1989; Juskevich and Guyer 1990) indicated that secondary, toxic effects are not significant. For example, bGH was not biologically active in humans, and oral toxicity studies have demonstrated that rabbit growth hormone (rbGH) was not orally active in rats, a species that is responsive to bGH (Westphal 1989). An increase in insulin-like growth factor (IGF-I) in the milk from cows with genetically transferred bGH has been observed. Oral toxicity studies have shown, however, that bovine IGF-I lacks oral activity in rats (Juskevich and Guyer 1990). Other studies have shown that concentration of IGF-I in milk of rbGH-treated cows is within the normal physiological range found in human breast milk. Furthermore, IGF-I is denatured under conditions used to process cow's milk for infant formula.

While biotechnology can increase agricultural productivity, produce greater profits, and lower food costs, thereby maintaining the competitiveness of U.S. agriculture in world markets, there are real and perceived adverse socioeconomic effects. Smaller dairy herds with lower per cow production levels are least likely to adopt the use of BST. The larger, higher producing dairy enterprises that do make use of BST will widen the production/profit gap with lower technology producers. This is likely to hasten the exit of many marginal operators from the industry. Likewise, use of PST to improve product quality in pork could lead to a restructuring of this industry. Thus, quality improvements and pressures to differentiate these new products in the retail marketplace could lead to changes in the market grades and/or the development of proprietary (branded) products. This trend would stimulate continued increased integration between producers and meat packers, resulting in a major restructuring of the production sector of the swine industry. Non-PST produced hogs might suffer in marketability under this scenario, causing those producers to exit the industry due to economic pressures. However, a significant consumer population will prefer products produced without the use of growth hormones, providing a limited market segment.

In spite of the bright prospects for the transgenic technology, industry willingness to invest in new biotechnologically derived products for plant and animal agriculture has dwindled noticeably (Shotwell 1991). The total research expenditure for BST by its several sponsors is now about $500 million. There is no real light at the end of the regulatory tunnel in either western Europe or the United States, and the future is uncertain so long as the public expresses a reluctance to accept the fruits of this technology.

There has been, however, a tremendous increase in fundamental knowledge. Other strategies to improve plants and animals may prove more acceptable to the public. Enormous effort has been made to understand lipid metabolism, but the primary maxim governing most thinking in agricultural circles is that the amount of body fat is little more than the net consequence of energy input and energy output. However, more detailed studies demonstrate an important role for neuroendocrine mechanisms in controlling fat stores in animals. In fish, the hormone prolactin is thought to have a central place in this mechanism (Horseman and Meier 1979).

Circadian Rhythm Control

Although porcine somatotropin (PST) decreases fat deposition, increases lean tissue deposition, and improves efficiency of food utilization, long-term administration of high doses of PST caused unsoundness (osteochrondosis) in pigs (Evock et al. 1988). An alternative approach for repartitioning body tissues in lambs, pigs, and laboratory species has involved a dopamine agonist, bromocriptine. Short-term bromocriptine administration reduced fat deposition in finishing pigs (Southern et al. 1990) and lambs (Eisemann et al. 1984). Bromocriptine inhibition of hepatic insulin receptor binding sites (Cincotta and Meier 1985b) suggests that the compound may mediate hepatic triglyceride synthesis and secretions (Southern et al. 1990). The compound may also disrupt the circadian rhythms that regulate lipogenesis (Cincotta and Meier 1985a). Thus, it may be possible to repartition body tissue compartments by simply altering circadian rhythms. Indeed, in laboratory animals (hamsters) daily one-hour temperature pulses above ambient caused significant increases in abdominal fat deposition (Waldrop and Meier 1985). Similar effects have been observed in fish subjected to brief exposures to warmed water (Spieler et al. 1977). Fish produced in culture and exposed to mechanical stimulation (sound) for 20 minutes each day demonstrated significant changes in body fat content. The amount of body fat varied depending on the time of daily sound stimulation. Some treatments had less fat than untreated controls, whereas others had more (Meier and Horseman 1977).

These studies suggest an interaction of a circadian stimulus rhythm with a circadian rhythm of target tissue responsiveness. Exploitation of these laboratory observations may well be beneficial to the production of farm animals (mammals, birds, and fish) without the public concerns for the unresolved ethical issues of biotechnology involving gene transfer or dietary hormonal treatments.

Environmental Taints

While carcass composition has a significant impact on muscle food quality, there are other influences on preharvest quality status. Most notable is off-flavor tainting. Animals can absorb compounds from the environment that impact the flavor of the meat. Grazing on specific plants can impart undesirable flavors to beef cattle. Another significant economic example is the uptake of off-flavors by farm-raised fish. Compounds such as geosmin and 2-methylisoborneol, two earthy/muddy off-flavors, are produced by microorganisms in the culture system, and are absorbed by the lipid tissues of the fish (Johnsen 1989). These plant metabolites give very strong flavors at very low concentrations (0.01 ppb). Uptake is rapid, and there is significant bioconcentration from the environment.

Biotechnology can play a significant role in eliminating the problem of environmental tainting and improving product quality. Efforts to establish the biosynthetic pathway for these compounds are underway (Naes, Utkilen, and Post 1988, 1989; Bently and Meganathan 1982; Dionigi et al. 1990). Metabolic regulators, such as mixed-function oxidase inhibitors, have been demonstrated to affect production of geosmin in a filamentous bacteria (Dionigi et al. 1990). Recent work suggests that these off-flavor metabolites may function to detoxify other metabolic compounds that are lethal to the organism when present at sufficient concentrations (Dionigi, Millie, and Johnsen 1991). Manipulation of the ability of specific organisms to utilize this internal biochemical defense strategy may permit the aquaculturalist to restructure the microbiological community to suppress or eliminate *only* off-flavor-producing species or strains. Thus, the growing environment, and not the consumer product, the fish, is involved in the genetic manipulation. Again this may reduce public concerns of the unknowns involved in the production of food using biotechnology.

IMPROVING MUSCLE FOOD QUALITY: POSTHARVEST

A major cause of quality deterioration in muscle foods is the rapid development of warmed-over flavor (WOF) following cooking. WOF was first recognized by Tims and Watts (1958) who defined it as the rapid onset of rancidity in cooked meats during refrigerated storage. Oxidized flavors are readily detectable after 48 hours in cooked meats, but the rancidity encountered in raw meats or fatty tissues develops more slowly and becomes evident only after prolonged freezer storage (Pearson and Gray 1983; Pearson, Love, and Shorland 1977). WOF can also develop rapidly in raw meat that has been ground and exposed to air (Greene 1969; Sato and Hegarty 1971). It is now generally accepted that any process that involves disruption of the muscle structure (e.g., cooking, grinding or restructuring) enhances the development of WOF. On the other hand, the acronym WOF is misleading, since it suggests that it is a flavor and not the dynamic process of flavor deterioration it has been shown to be (Johnsen and Civille 1986; St. Angelo et al. 1987, 1988; Spanier, Edwards, and Dupuy 1988; Spanier, McMillin, and Miller 1990; Drumm and Spanier 1991). MFD, or meat flavor deterioration, has been proposed as a more accurate acronym to describe the dynamic changes in meat flavor that accompany storage (Spanier, McMillin, and Miller 1990; Drumm and Spanier, 1991). During MFD, the undesirable flavors increase in intensity, while the desirable meaty flavors decline (Spanier, Vercellotti and James 1991).

Many commercially available compounds retard the deterioration of meat flavor. These compounds usually protect flavor via an antioxidant or chelator mechanism. Some of these products are not true antioxidants, being Maillard reaction products (MRP) or reductones that can act as antioxidants (Bailey et al. 1987). From the perspective of food flavor it is unfortunate that many MRPs with antioxidative activity also have distinctive taste(s). One such compound with effective antioxidant and MFD-retarding activity is maltol (St. Angelo et al. 1988). However, maltol, a major flavor component of cotton candy, gives the ground beef patty the flavor of this confectionery. The drawback to several of the other antioxidants and chelators is many are salts that provide a food product excellent shelflife but exacerbate hypertension in susceptible recipients.

N-Carboxymethylchitosan

A new compound that has been demonstrated to be a highly effective retardant of MFD has the acronym NCMC for N-carboxymethylchitosan (St. Angelo et al. 1988). The structure of NCMC is shown in Figure 10–1. It is believed that this compound functions through its iron-chelating ability, that is, NCMC inhibits the initiation of lipid oxidation and development of rancidity induced by metal catalysis. A patent entitled "Inhibition of Warmed-over Flavor and Preserving of Uncured Meat Containing Materials" (U.S. Patent Number 4,871,556; Oct. 3, 1989) has been issued for the use of NCMC. Although applicable to meat/muscle foods, it could be applied to other food types as well. NCMC is derived from a natural product, the carapace of crustaceans, such as crabs and shrimp, and is therefore

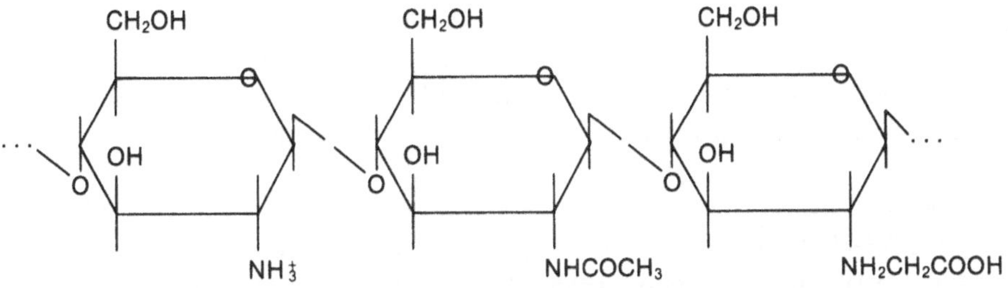

FIGURE 10-1.   The structure of *N*-carboxymethylchitosan.

more acceptable to today's consumer. NCMC imparts no additional taste. The compound is highly effective at low concentrations, that is, 5000 ppm (0.5%), and is soluble in water, making it a convenient product for addition to many food items. Because NCMC is a polysaccharide, it is hypoallergenic; and because it is not a salt, there is no danger that it could cause or exacerbate existing hypertension. NCMC should prove valuable as a food additive and shelf-life extender in the years to come.

Value-Added Strategies

The quality of muscle food has been enhanced by the addition of various products derived from bacteria and other microbes. For example, beefy-, chickeny-, and fishy-flavored Maillard reaction products are formed after appropriate heating of various strains of yeast grown in large, commercial bioreactor vats. Similarly, many other food enzymes and food flavorants have been obtained from chemically and/or biologically modified microorganisms and from organisms whose growth has been regulated by some recently developed engineering processes. An excellent text covering much of this subject matter recently appeared under the title *Biotechnology and Food Process Engineering* (Schwartzberg and Rao 1990). Tremendous progress has also been made in the development of the technology of bioengineered plants, bacteria, and fungi (Gasser and Fraley 1989; Lindow, Panopoulos, and McFarland 1989; Timberlake and Marshall 1989). The relative importance to muscle food quality of genetically engineered plant and microbial products lies in the use of the products as flavor enhancers or adjuvants. Some pigments, many proteins, and a variety of enzymes, carbohydrates, and other nutrients that are either generally recognized as safe (GRAS) or subject to approved food additive regulations can be produced by modern biotechnology fermentation methods. However, when site specific genetic insertions in the seed stock are made, the products lose their GRAS status (Shotwell 1991).

A potential route for the utilization of this technology, as it relates to muscle foods, resides in the genetic engineering of flavor quality proteins and peptides. This could include, and would not be restricted to, the production of specific proteinases, such as the calcium-activated papain-like enzymes and the cathepsins. The calcium-activated papain-like enzymes, referred to in the literature under several names, that is, calpains, CDP-I and CDP-II, CAF, and CANP, would be used in producing more tender beef cuts (Koohmaraie et al. 1988; St. Angelo et al. 1991). The cathepsins, particularly the thiol proteinases, cathepsins B, H, and L, would be used for stimulating the production of natural-meaty-flavor compounds and precursors (Spanier, McMillin, and Miller 1990). If there is a drawback to this methodology, it would be the possibility that the flavor-enhancing enzyme, protein, or peptide could be hyperallergenic. The concern over hyperallergenicity is reduced if the flavor protein or peptide is found as a natural product of muscle. Native proteins are traditionally nonallergenic. Further, in the case of most of the hormones found in transgenic animals, the potential allergenicity to the dining host is minimized through the protein's hydrolysis in the stomach and intestines (Westphal 1989; Juskevich and Guyer 1990).

A natural small, linear oligopeptide that contains eight amino acid residues has been identified (Yamasaki and Maekawa 1978). This peptide has a beefy and meaty taste, and has been named *BMP* (Spanier et al. 1992). It is possible to envision the genetic engineering of BMP and related peptides into some vegetable crop or microbe for later isolation and use as a flavor adjuvant. Further, additional modification(s) could be made to the structure of BMP that might increase its usefulness, for example, make it highly resistant to temperature denaturation, to proteolytic fragmentation, or make it meatier tasting.

CONCLUSION

Technology is becoming available for the production or enhancement of higher quality muscle foods. Results from the mechanistic approach to meat flavor research will permit us to develop better predictive, adaptive and/or management methods for enhancing meat flavor quality. However, significant impediments to implementing this knowledge are found in the socioethical concerns aroused in the public when the food supply is being manipulated by these new techniques. The main thrust of future research should be directed toward the discovery of more natural flavor constituents so muscle food products of high nutritional and flavor value can meet the concerns of tomorrow's consumers about how these products were produced.

References

Bailey, M. E. 1988. Inhibition of warmed-over flavor, with emphasis on Maillard reaction products. *Food Technol.* **42**:123–126.

Bailey, M. E., S. Y. Shin-Lee, H. P. Dupuy, A. J. St. Angelo, and J. R. Vercellotti. 1987. Inhibition of warmed-over flavor by Maillard reaction products. In *Warmed-over Flavor of Meat,* ed. A. J. St. Angelo and M. E. Bailey, pp. 237–266. Orlando, FL: Academic Press.

Bentley, R., and R. Meganathan. 1982. Geosmin and methylisoborneol biosynthesis in Streptomyces, evidence for a isoprenoid pathway and its nondifferentiating isolates. *FEBS Lett.* **3**:220–222.

Cincotta, A. H., and A. H. Meier. 1985a. Prolactin permits the expression of a circadian variation in lipogenic responsiveness to insulin in hepatocytes of the golden hamster (*Mesocricetus auratus*). *J. Endocrinol.* **106**:173–176.

Cincotta, A. H., and A. H. Meier. 1985b. Prolactin permits the expression of a circadian variation in receptor profile in hepatocytes of the golden hamster (*Mesocricetus auratus*). *J. Endocrinol.* **106**:177–181.

Dionigi, C. P., D. F. Millie, and P. B. Johnsen. 1991. Effects of farnesol and the "off-flavor" derivative geosmin on *Streptomyces tendae. Appl. Environ. Microbiol.* **57**:3429–3432.

Dionigi, C. P., D. A. Greene, D. F. Millie, and P. B. Johnsen. 1990. Mixed function oxidase inhibitors affect production of the off-flavor microbial metabolite geosmin. *Pest. Biochem. Physiol.* **38**:76–80.

Drumm, T. D., and A. M. Spanier. 1991. Changes in the content of lipid autoxidation and sulfur-containing compounds in cooked beef during storage. *J. Agric. Food Chem.* **39**:336–343.

Eisemann, J. H., D. E. Bauman, D. E. Hohue, and H. F. Travis. 1984. Influence of photoperiod and prolactin on body composition and *in vitro* lipid metabolism in wether lambs. *J. Anim. Sci.* **59**:95–104.

Etherington, D. J., M. A. J. Taylor, and E. Dansfield. 1987. Conditioning of meat from different species. Relationship between tenderizing and the levels of cathepsin B, cathepsin L, calpain I, calpain II and β-glucuronidase. *Meat Sci.* **20**:1–18.

Evock, C. M., T. D. Etherton, C. S. Chung and R. E. Ivy. 1988. Pituitary porcine growth hormone (pGH) and recombinant pGH analog stimulate pig growth performance in a similar manner. *J. Anim. Sci.* **66**:1928–1941.

Fallert, R., T. McGuckin, C. Betts, and G. Bruner. 1987. "bST and the Dairy Industry: A National, Regional, and Farm-level Analysis." U.S. Department of Agriculture, Economic Research Service, Agricultural Economic Report No. 579.

Gasser, C. S., and R. T. Fraley. 1989. Genetically engineering plants for crop improvement. *Science* **244**:1293–1299.

Greene, B. E. 1969. Lipid oxidation and pigment changes in raw beef. *J. Food Sci.* **34**:110–113.

Hayenga, M. L. 1988. "Biotechnology in the Food and Agricultural Sector: Issues and Implementations for the 1990s." Agricultural Issues Center, University of California, Davis, UC AIC Issues Paper No. 88-5.

Horseman, N. D., and A. H. Meier. 1979. Circadian phase-dependent prolactin mechanisms in hepatic lipogenesis of a telost. *J. Endocrinol.* **82**:367–372.

Johnsen, P. B. 1989. Factors influencing the flavor quality of farm-raised catfish. *Food Technol.* **43**:94–97.

Johnsen, P. B., and G. V. Civille. 1986. A standardized lexicon of meat WOF descriptors. *J. Sensory Studies* **1**:99–104.

Juskevich, J. C., and C. G. Guyer. 1990. Bovine growth hormone: Human food safety evaluation. *Science* **249**:875–884.

Kapuscinski, A. R., and E. M. Hallerman. 1990a. Transgenic fish and public policy: Anticipating environmental impacts of transgenic fish. *Fisheries* **15**:2–11.

Kapuscinski, A. R., and E. M. Hallerman. 1990b. Transgenic fishes: AFS position statement. *Fisheries* **15**:2–5.

Kato, H., and T. Nishimura. 1987. Taste components and conditioning of beef, pork, and chicken. In *Umami: A Basic Taste,* ed. Y. Kawamura and M. R. Kare, pp. 289–306. New York: Marcel Dekker.

Koohmaraie, M., A. S. Babiker, A. L. Schroeder, R. A. Merkel and T. R. Dutson. 1988. Acceleration of postmortem tenderization in ovine carcasses through activation of Ca^{2+}-dependent proteases. *J. Food Sci.* **53**:1638–1641.

Lindow, S. E., N. J. Panopoulos, B. L. McFarland. 1989. Genetic engineering of bacteria from managed and natural habitats. *Science* **244**:1300–1307.

Meier, A. H., and N. D. Horseman. 1977. Stimulation and depression of growth, fat storage, and gonad weight in the teleost fish, *Tilapia aurea.* In *Proceedings of the 8th Meeting of the World Mariculture Society,* ed. J. Avault, pp. 135–143. San Jose, Costa Rica: Louisiana State University, Division of Continuing Education.

Naes, H., C. Utkilen, and A. F. Post. 1988. Factors influencing geosmin production by the cyanobacterium *Oscillatoria brevis. Water Sci. Technol.* **20**:125–131.

Naes, H., C. Utkilen, and A. F. Post. 1989. Geosmin production in the cyanobacterium *Oscillatoria brevis. Arch. Microbiol.* **151**:407–410.

Ouali, A. 1990. Meat tenderization: Possible causes and mechanisms. A review. *J. Muscle Foods* **1**:129–165.

Ouali, A., N. Garrel, A. Obled, C. Deval, C. Valin, and I. F. Penny. 1987. Comparative action of cathepsins D, B, H, L and of a new lysosomal cysteine proteinase on rabbit myofibrils. *Meat Sci.* **19**(2):83–100.

Pearson, A. M., and J. I. Gray. 1983. Mechanism responsible for warmed-over flavor in cooked meat. In *The Maillard Reaction in Foods and Nutrition.* American Chemical Society Symposium Series 215, ed. G. R. Waller and M. S. Feather, pp. 287–300. Washington, DC: American Chemical Society.

Pearson, A. M., J. D. Love, and F. B. Shorland. 1977. "Warmed-over" flavor in meat, poultry, and fish. In *Advances in Food Research,* Vol. 23, ed. C. O. Chichester, E. M. Mrak, and G. F. Stewart, pp. 1–74. New York: Academic Press.

Pursel, V. G., C. E. Rexroad, Jr., D. J. Bolt, K. F. Miller, R. J. Wall, R. E. Hammer, C. A. Pinkert, R. D. Palmiter, and R. L. Brinster. 1987. Progress on gene transfer in farm animals. *Vet. Immunol. Immunopathol.* **17**:303–312.

Pursel, V. G., C. A. Pinkert, K. F. Miller, D. J. Bolt, R. G. Campbell, R. D. Palmiter, R. L. Brinster, and R. E. Hammer. 1989. Genetic engineering of livestock. *Science* **244**:1281–1299.

St. Angelo, A. J., J. R. Vercellotti, M. G. Legendre, C. H. Vinnett, J. W. Kuan,

C. James, and H. P. Dupuy. 1987. Chemical and instrumental analyses of warmed-over flavor in beef. *J. Food Sci.* **52**:1163–1168.

St. Angelo, J. R. Vercellotti, H. P. Dupuy, and A. M. Spanier. 1988. Assessment of beef flavor quality: Multidisciplinary approach. *Food Technol.* **42**:133–138.

St. Angelo, A. J., M. Koohmaraie, K. L. Crippen, and J. Crouse. 1991. Simultaneous acceleration of postmortem tenderization and inhibition of warmed-over flavor by infusion of calcium chloride plus antioxidants into lamb carcasses. *J. Food Sci.* **56**:359–362.

Sato, K., and G. R. Hegarty. 1971. Warmed-over flavor in cooked meats. *J. Food Sci.* **36**:1098–1102.

Schwartzberg, H. G., and M. A. Rao. 1990. *Biotechnology and Food Process Engineering.* New York: Marcel Dekker.

Shotwell, T. K. 1991. Regulatory outlook for biotechnology, direct-fed microbials not bright. *Feedstuffs* **63**:26, 52.

Southern, L. L., A. H. Cinotta, A. H. Meier, T. D. Bidner, and K. L. Watkins. 1990. Bromocriptine-induced reduction of body fat in pigs. *J. Anim. Sci.* **68**:931–936.

Spanier, A. M., J. V. Edwards, and H. P. Dupuy. 1988. The warmed-over flavor process in beef: A study of meat proteins and peptides. *Food Technol.* **42**:110–118.

Spanier, A. M., K. W. McMillin, and J. A. Miller. 1990. Enzyme activity levels in beef: Effect of postmortem aging and end-point cooking temperature. *J. Food Sci.* **55**:318–326.

Spanier, A. M., J. R. Vercellotti, and C. James, Jr. 1991. Mapping beef flavor quality: Correlation of sensory, instrumental, and chemical attributes as influenced by structure and oxygen exclusion. *J. Food Sci.* **57**:10–15.

Spanier, A. M., J. A. Miller, and J. M. Bland. 1992. Lipid oxidation: Effect on meat proteins. In *Lipid Oxidation in Food,* A. J. St. Angelo, ed. Washington, DC: ACS Books (in press).

Spieler, R. E., T. A. Noeske, V. deVlaming, and A. H. Meier. 1977. Effects of thermocycles on body weight gain and gonadal growth in the goldfish, *Carassius auratus. Trans. Am. Fish. Soc.* **106**:440–444.

Timberlake, W. E., and M. A. Marshall. 1989. Genetic engineering of filamentous fungi. *Science* **244**:1313–1317.

Tims, M. J., and B. M. Watts. 1958. Protection of cooked meats with phosphates. *Food Technol.* **12**:240–243.

Waldrop, R. D., and A. H. Meier. 1985. Fattening responses to daily intervals of heat exposure in the Syrian hamster. *Life Sci.* **37**:1539–1543.

Westphal, H. 1989. Transgenic mammals and biotechnology. *FASEB J.* **3**:117–120.

Yamasaki, Y., and K. Maekawa. 1978. A peptide with delicious taste. *Agric. Biol. Chem.* **42**:1761–1765.

Index

Acetaldehyde, 76
Actinomycetes, 114, 118, 127
Adenosine triphosphate, continuous
    generation of, 54
Aflatoxin biosynthetic pathway, 206,
    216
  enzymes in, 218
  genes in, 219
  inhibitors, interruption of, 216, 220
  intermediates in, 206, 216
  regulation, 219
Aflatoxins, 205, 207, 208
Agriculture, 230
Agrobacterium tumefaciens, 9, 25, 216
Albedo, 70, 75
Albumins, 5
Alpha-amylase, 86, 88, 90
  maltogenic, 86, 93
Amino acids, essential, 1, 4
Amphilicity, 13
Amplification techniques. *See*
    Polymerase chain reaction
Animal, 230
Anions, catalysts in chemical
    deamidation, 40
Antibiotic, 114, 132
  production, 90
  resistance markers, 87
    aminoglycoside, 91
    bleomycin, 89
    FDA concerns, 87
    kanamycin, 89
    phleomycin, 93
Antibody, antiserium and ELISA
    probes, 106. *See also Listeria*,
    detection methods
Antioxidant, 236
Apple, 76
Aquaculture, 230, 235, 236
Aspergillus spp.
  alliaceus, 117
  flavus, 77, 205
  japonicum, 66
  niger, 86, 92
  nonaflatoxigenic, 209, 220
  parasiticus, 205
  versicolor, 77
Association of Official Analytical

Chemists (AOAC) test kits. *See*
    Listeria, detection methods
Automated equipment. *See* Robot
Automation. *See* Robot

Bacillus spp.
  cereus, 151, 158, 165
  megaterium, 86, 90
  stearothermophilus, 86, 89, 93
  subtilis, 86, 93
Bacteria, 86, 114, 118, 121, 127, 151
Bacterial pathogens
  early detection, 190
  foodborne, 189
Banana, 61
Benzaldehyde, 76
Bioactive peptides, 233
Biocompetitive agents,
    nonaflatoxigenic, 209, 220
Bioreactor, 237
Biosynthesis
  aflatoxin, 206, 216
  flavor compounds, 235
BMP, 238
Body fat, 234
Bovine growth hormone, 230
Breeding
  animal, 229
  microbial, 113
  plant, 4, 205
Bromocriptine, 234

Calcium, 70, 75
  activated enzyme, 238
Campylobacter jejuni, 151, 159, 165
Capillary membrane. *See* Segment
    membrane
Carapace, 236
Carboxykinase, 230
Carcass, 235
Cardiomegaly, 232
Casein kinase, 53. *See also* Protein
    phosphorylation
Cathepsins, 238
Cation exchange resin, catalyst in
    chemical deamidation, 40
Cellulose, 66, 73
Chelator, 236

Chromobacterium, 115
  genetic engineering, 113, 124
  inhibitor, 115
  isolation, 121, 124, 129, 141, 146
Chymosin, 88
  from *Aspergillus niger*, 86, 92
  from *Escherichia coli*, 86, 90
  from *Kluyveromyces marxianus*, 86,
    91
Circardian rhythm, 234
Citrus. *See* Nonclimacteric fruit
Climacteric fruit
  apple, 76
  banana, 61
  description of, 61
  mango, 77
  papaya, 63
  peach, 63, 70
  pear, 63, 77
  tomato, 61
Clostridium spp.
  botulinum, 151, 156, 170
  perfringens, 151, 158, 165, 170
Coumarin, 71
CPC International, Inc., 86, 90

Dairy, 233
Deamidation. *See* Protein deamidation
Deficiency values, 16
Digestibility, 24
DNA probes
  cell viability, 156
  development, 152
  formats. *See* Hybridization
  labels, 155
  sensitivity, 154
  specificity, 155
  target. *See* Plasmids; Ribosomal
    RNA; Toxins
Dopamine, 234

ELISA, 197
Endopolygalacturonase. *See*
    Polygalacturonase
Environment, 87, 232
  assessment of, 87, 93
  containment, 88
  fate of organisms in, 88
  National Environmental Policy Act,
    88

Enzymatic phosphorylation, 47
Enzyme Bio-Systems Ltd., 86, 90. *See
    also* Protein phosphorylation
Enzymes
  α-amylase, 89
  chitinase, 104, 211
  chymosin, 88
  desaturase, 218
  esterase, 218
  glucanase, 104, 211
  β-glucosidase, 71
  history of food use, 86, 90
  hydrolase, 63
  methyltransferase, 218
  naringinase, 70
  oxidoreductase, 218
  pectolyase, 66
  reductase, 218
Escherichia coli, 89, 151, 158, 165, 170
  enterohaemorrhagic, 161
  enteroinvasive, 161, 163
  enteropathogenic, 161
  enterotoxigenic, 155, 161
  K-12, 89, 91
Essential amino acid, 1, 4
Ethylene, 61, 64
Exomaltohydrolase. *See* Alpha-
    amylase, maltogenic
Expolygalacturonase. *See*
    Polygalacturonase

Feed efficiency, 232
Fermentation, 237
Flagella extracts. *See Listeria*,
    detection methods
Food, 84, 94, 229
  additives, 84
  confectionery, 236
  cooking, 229, 235
  derived from microorganisms, 95
  flavor, 229, 235, 238
  flavorants, 237
  generally recognized as safe (GRAS),
    237
  marketplace, 233
  meat, 229, 235
  meat flavor deterioration (MFD), 236
  mixtures, 94
  retail, 233
  storage, 229, 235

Food (*continued*)
  value-added, 237
  whole, 94
Foodborne illness/outbreak, 151, 157, 165, 189
Foodborne pathogenic bacteria, 189
Food Chemical Codex, 87, 90
Food and Drug Administration, 84
Food, Drug and Cosmetic Act, 84
Fructose, 66
Fungi, 66, 76, 92, 117, 209, 237

Galactose, 66
Galacturonic acid. *See* Pectin
Genencor, Inc., 86, 92
Generally recognized as safe. *See* GRAS
Geosmin, 235
Gist-Brocades, Inc., 86, 91
Globulins, 5
Glucose, 66, 71
β-Glucosidase, 71
Glutelins, 5
Glycinin, 5
Good Industrial Large Scale Practice (GILSP), 92
Good Laboratory Practices, 87
GRAS, 84, 85, 95, 237
Growth factor, 233

Hazard analysis critical control points (HACCP), 190
Health
  arthritis, 232
  cancer, 205
  dermatitis, 232
  gastric ulcers, 232
  hepatic, 234
  hyperallergenic, 238
  hypertension, 236
  hypoallergenic, 237
  insulin, 233
  osteochrondosis, 234
  renal disease, 232
Hemolysin. *See* Listeria, detection methods
Host-plant resistance genes
  against aflatoxin production, 207, 210
  plant breeding vs. genetic engineering, 207, 210

for plant chemicals, 210
plant transformation, 211, 215
tissue specific expression, 213
Hybridization, 152
  colony, 153, 159, 161, 166
  liquid, 153, 157
Hydrolase
  carboxylester, 63. *See also* Pectin methylesterase
  random, 63. *See also* Polygalacturonase

IIT Research Institute, 94
Illinois Institute of Technology, 94
Immuno-dot-blot. *See* Listeria, detection methods
International Food Biotechnology Council, 95
Iturin, 77

Joint FAO/WHO Expert Committee on Food Additives, 95

Kanamycin segregation pattern, 27
Kluyveromyces marxianus, 86, 91
Kwashiorkor, 3

Lactose, 66
Limonin, 71
Lipid oxidation, 229, 236
Lipogenesis, 234
Listeria monocytogenes, 151, 158
  detection methods, 190
  ecology of, 189
  rapid detection methods
    nucleic acid probes, 191
    antibodies, selological probes, ELISA, 191, 196
    polymerase chain reaction (PCR), 195
    unique listeriolysin O gene probe, 195
Listeriosis, 189
Listerolysin O gene. *See* Listeria, detection methods

Maillard reactions, 229
Maillard reaction product (MRP), 236
Maltol, 236
Mango, 77
Mannose, 66

Metabolic function, 232, 235
Metabolism, 229, 233
Metallothionein, 230
Methylisoborneol, 235
Microorganisms
 bacteria, 9, 84, 115, 151, 232, 235
 fungi, 66, 77, 92, 104, 117, 205, 220, 237
 virus, 169
 yeast, 237
Middle lamella, 63, 72
Mixed-function oxidase inhibitors, 235
Monascus, 117
 mutant, 122, 136
Muscle, 229, 235

Naringenin, 71
Naringin, 70, 75
Naringinase, 70
National Center for Toxicological Research, 94
National Institutes of Health (NIH) guidelines, 89
National Laboratory for Food Safety and Technology, 94
N-carboxymethylchitosan (NCMC), 236
Needle, 124, 129, 133, 136
Neuroendocrine, 234
Neurospora crassa, 92
Nonclimacteric fruit, 61
 citrus, 61, 68, 73, 77
 grapefruit, 70, 74
 orange, 68, 72, 76
Novo Laboratories, Inc., 86, 93

Off-flavor, 235
Office of Biotechnology, 93
Office of Science and Technology Policy, 94
Oligouronide. *See* Pectin
Oral toxicity, 90, 93, 233
Orange, 68, 72, 76
Oxygen barrier film, 75, 76

Panicoidae, 5
Papaya, 63
Pathogenicity, 90, 92, 172, 189
Peach, 63, 70
Pear, 63, 77
Pectin, 63, 67, 69
 galacturonic acid, 63, 69

polypectase, 63, 65
 simple oligouronides, 63, 67
 unsaturated oligouronide, 63, 67
Pectin lyase (endo-PL), 63, 66, 72
Pectin methylesterase (PME), 63, 70, 72
Pectolyase, 66, 68
Penicillium spp., 117
 digitatum, 76
 italicum, 76
P-enolpyruvate carboxykinase, 230
Pfizer, Inc., 86, 91
Phosphorus oxychloride. *See* Protein phosphorylation, chemical
Phosphorylation. *See* Protein phosphorylation
Phytoalexin, 64, 211
Plants
 breeding, 4, 205, 207, 210
 genetic engineering, 9, 210
 resistance genes, 100, 210
 transformation, 9, 211, 215
Plasmids, 158, 161, 164, 167
 pBR322, 91
 pBR327, 89
 pUB110, 89, 93
 pUC18, 91
Polygalacturonase (PG), 72
 endopolygalacturonase (endo-PG), 63, 214
 exopolygalacturonase (exo-PG), 63, 70
Polymerase chain reaction, 156, 166, 195
Polypectate, 63, 69
Polysaccharide, 237
Potassium sorbate, 75
Prolamines, 5
Protein(s), 1, 2
 de novo design of, 2, 12, 20
  long-range stabilizing interactions, 13
  short-range stabilizing interactions, 12
 methionine-rich, 11
 quality, 24
 requirement, 25
 storage, 1, 7, 10
 vegetative storage, 8
Proteinases, 238
 inhibitors, 214

Protein bodies, 5
Protein deamidation, 38
  chemical, 40
  catalyzed by anions, 40
  enzymatic, 41
    by peptidoglutaminase, 44
    by protease, 41
    by transglutaminase, 42
Protein kinases, 50. *See also* Protein
    phosphorylation
Protein phosphorylation, 47
  chemical, 47
  enzymatic, 50, 53
Prunin, 71
Pyrrolnitrin, 77

Raffinose, 66
Rancidity, 235
Rhamnose, 66
Recombinant DNA safety evaluation
  effect on GRAS status, 86
  genetic exchange, 87
  organism description, 87
  research, 94
  safety information for petitions, 87.
    See also Toxicological testing
Reductones, 236
Ribosomal RNA, 152, 159, 163, 170,
    192
Robot, 112, 123, 127, 131, 140, 148

Saccharomyces cerevisiae, 91
Salmonella, 151, 158, 162, 165, 170
Screening, microbial, 114
Segment membrane, 70, 76
  capillary membrane, 72
Serotypes, *Listeria*, 192
Shigella, 152, 163, 170
Sodium trimetaphosphate, 47
Somatotropin, 232
Sorbitol, 66
Staphlococcus aureus, 152, 165
Streptomyces, 115
Subcutaneous fat, 232
Sucrose, 66, 71
Sugars, neutral, 66, 71

Tolypocladium, 115
Toxicological testing, 95. *See also*
    Toxins

acute oral toxicity, 90
allergenicity, 92
cell toxicity, 92
dermal irritation, 93
eye irritation, 93
inhalation toxicity, 93
mutagenicity, 92
oral gavage, 91
palatability, 90
sensitization, 93
subacute oral toxicity, 93
subchronic oral toxicity, 90
Toxins, 152, 158, 170
  aflatoxins, 205
  mycotoxins, 87, 92, 205
  Shiga-like, 89, 91
  Staphylococcus aureus enterotoxins,
    89
Trans-eliminase. *See* Pectin lyase
Transgenic, 7, 230, 238
Triglyceride, 234
Triticeae, 5

University of Illinois, 94

Vacuum-infusion, 64, 71
Vibrio
  cholerae, 152, 165, 170
  parahaemolyticus, 152, 165, 170
  vulnificus, 152, 166, 171
Viruses
  coxsackievirus, 172
  echovirus, 173
  enterovirus, 169, 173
  hepatitis A, 171
  Norwalk, 171
  picornavirus, 172
  poliovirus, 172
  rotavirus, 172

Warmed-over flavor (WOF), 235
Weight gain, 232

Xylose, 66

Yersinia
  enterocolitica, 152, 167
  pseudotuberculosis, 152, 167

Zein, 6, 11